WEAPON

HOTCHKISS
MACHINE GUNS

JOHN WALTER
Series Editor Martin Pegler

Illustrated by Adam Hook & Alan Gilliland

OSPREY PUBLISHING
Bloomsbury Publishing Plc
PO Box 883, Oxford, OX1 9PL, UK
1385 Broadway, 5th Floor, New York, NY 10018, USA
E-mail: info@ospreypublishing.com
www.ospreypublishing.com

OSPREY is a trademark of Osprey Publishing Ltd

First published in Great Britain in 2019

© Osprey Publishing Ltd, 2019

A catalogue record for this book is available from the British Library.

ISBN: PB 9781472836168; eBook 9781472836151; ePDF 9781472836144; XML 9781472836175

19 20 21 22 23   10 9 8 7 6 5 4 3 2 1

Index by Rob Munro
Typeset by PDQ Digital Media Solutions, Bungay, UK
Printed and bound in India by Replika Press Private Ltd.

Osprey Publishing supports the Woodland Trust, the UK's leading woodland conservation charity.

To find out more about our authors and books visit www.ospreypublishing.com. Here you will find extracts, author interviews, details of forthcoming events and the option to sign up for our newsletter.

Dedication

To Alison, Adam, Nicky, Findlay, Georgia and Holly – not forgetting Amber the Dog – for help and encouragement.

Acknowledgements

I would particularly like to thank Lisa Oakes, formerly of James D. Julia, and Sarah Stoltzfus of Morphy's Auctions for providing such excellent images. I'd also like to mention Alexander and Christian Cranmer of International Military Antiques Inc.; Cédric and Julie Vaubourg of Fortiff' Séré; Rock Island Auctions; Cowan's Auctions; *American Rifleman* magazine; Kongsberg Våpenhistoriske Forening; and the US National Archives. The websites of British, French, German and US patent offices gave access to specifications and illustrations. E-addresses appear, where appropriate, with individual captions. I'd also like to thank Paul Scarlata and Ian Skennerton for information concerning the Benét-Mercié and the British-service Hotchkisses respectively; and Osprey's Nick Reynolds and the series editor Martin Pegler for not only believing in the project, but also for the suggestions that have improved coverage.

Artist's note

Readers may care to note that the original paintings from which the colour plates in this book were prepared are available for private sale. All reproduction copyright whatsoever is retained by the publishers. All enquiries should be addressed to:

Scorpio, 158 Mill Road, Hailsham, East Sussex BN27 2SH, UK
Email: scorpiopaintings@btinternet.com

The publishers regret that they can enter into no correspondence upon this matter.

Front cover, above: The 13.2×99mm Hotchkiss was introduced in 1930 to compete with the legendary .50 Browning. However, the length of the French cartridge case so nearly matched that of the Browning round (12.7×99mm) that if the two types of gun were being used in the same theatre, attempts would be made to load the wrong cartridges. This had a potential to cause a breech explosion, and so Hotchkiss substituted a 13.2×96mm cartridge in 1935. (© Royal Armouries PR.1760)

Front cover, below: The Taishō 3rd Year Type machine gun in service with the Imperial Japanese Navy's Special Landing Forces. Note the spade-type handgrips, the easiest way of distinguishing the 3rd Year Type from its successor, the Type 92 heavy machine gun. (© CORBIS/Corbis via Getty Images)

Title-page photograph: The son of Italian immigrants, Staff Sergeant Frank Iacono is pictured on 27 March 1944 while training at San Luis Obispo, California, with a captured Japanese Type 92 heavy machine gun on a US M1917 tripod. (US Army Signal Corps)

CONTENTS

INTRODUCTION

The Hotchkiss machine gun, scarcely remembered today, not only successfully pioneered gas operation but was one of the first of its type to be exported in quantity. Introduced commercially just as the 20th century dawned, the Hotchkiss achieved the first 'kill' in aerial combat over the Marne in 1914, became the first machine gun serving the American Expeditionary Forces to fire shots in anger in 1917, and armed both sides during the Spanish Civil War of 1936–39. Many of its derivatives served during World War II – the most notorious, perhaps, being the Japanese adaptations that extracted a terrible toll as US armed forces advanced island-by-island across the Pacific Ocean. The bitter and intense fighting on Guadalcanal, Tarawa, the Marianas and Iwo Jima, has passed into folklore.

The Hotchkiss is also an important link in a chain stretching back hundreds of years. The multi-shot gun, and also, perhaps surprisingly, the breechloader, had been created not so long after the handgonne (hand cannon) had appeared in Europe in the early 13th century. Medieval guns survive with several barrels which were hammer-welded together in a cluster, sometimes to be fired individually – difficulties of priming notwithstanding – or else as a volley.

The ribauldequin, *Orgelgeschütz* or volley gun consisted of a considerable number of barrels, placed either in a horizontal row or sometimes in several superimposed rows, which were fired by a single train of powder in an ignition channel to the rear of the barrels. Flash, noise and smoke may have been diabolic, but the effects, once the shock of novelty had dissipated, fell far short of what was desired. Consequently, the multi-shot idea lay dormant until technology offered new solutions. Magazine firearms such as the Scandinavian Kalthoff and the Italian Lorenzoni appeared in the 17th century, and, by the 1780s, the British were giving consideration to Jover & Belton muskets with superimposed charges and a flintlock which slid longitudinally to fire each charge individually. In reality, however, the difficulties of sealing the charges proved too much of a handicap and chain-firing was by no means uncommon.

The volley gun reappeared, though the number of shots was still restricted by the weight of the barrels. Yet inventors refused to relent. In August 1824, the American-born engineer Jacob Perkins patented a method of firing projectiles by steam generated in his high-pressure boiler. Trials reported in *The London Mechanic's Register* on 6 November 1824 showed the gun to work surprisingly well, flattening lead balls against a steel plate placed 100ft (30m) away and capable of firing 1,000rd/min. So effective was the Perkins Gun that the Commander-in-Chief of the British Army, the Duke of Wellington, refused even to consider it as a weapon on the grounds that it was far too destructive. In addition, the high-pressure boiler was justifiably reckoned to be a liability in battle.

The use of multi-shot weapons in combat had to wait for the American Civil War (1861–65). The unique circumstances of this conflict created an environment in which, largely owing to shortages of regulation weapons, inventors and entrepreneurs alike were given the chance to prove their ideas in combat. Alongside a multitude of cap-lock revolvers were the Spencer and Henry repeating rifles, and many single-shot designs chambering the self-contained metal-case ammunition that pointed to the future. From this era also came the multi-barrel, rapid-fire Gatling Gun.

Benjamin Berkeley Hotchkiss produced shells and fuzes during the American Civil War, initially by exploiting the ideas of his late brother Andrew, and made a considerable fortune; but he had seen at first hand the advantages of not only self-contained ammunition but also the potential that lay in the Gatling Gun.

When war between France and Prussia began in July 1870 and the fighting went against the French, huge numbers of weapons were imported to arm not only the Garde Nationale reserve force but also the irregulars of the Gardes Mobiles who continued to fight long after much of the French Army had surrendered.

A 1909-type Benét-Mercié Machine Rifle being fired by Hiram Percy Maxim (1869–1936), son of Hiram and the inventor of the silencer shown under test. The 1914 *Annual Report of the Chief of Ordnance* noted that two silencers had been tested. Differing in the number of chambers – ten in Model A, 12 in Model B – neither was entirely satisfactory. However, the testers recommended acquiring four examples of a strengthened Model A. There is no evidence that these trials were ever carried out, however, as a rapidly firing machine-gun such as the M1909 will soon burn-out silencers of this type. (Author's Collection)

The crew of a Japanese Taishō 11th Year Type machine gun take aim. Note the large canvas bag being held next to the ejection port to catch spent cases – suggesting a training exercise – and the rudimentary camouflage on the men's caps. The photograph gives a good impression of the manoeuvrability of the 11th Year Type, but also shows the vulnerability of the feed hopper, visible on the left side of the breech, to ingress of dust and debris. The accompanying infantrymen, who would probably not be standing so prominently if combat was imminent, carry 6.5mm 38th Year Type Arisaka rifles with sword-bayonets fixed to their muzzles. (Author's Collection)

Like many inventors, Hotchkiss believed that the Dreyse and Chassepot needle-rifles were developmental dead ends; and the same could be said of the vaunted 25-barrel 13mm de Reffye Mitrailleuse that the French considered not only to be a state secret but also the arbiter of battle. They were mistaken: volley firing, even if pre-loaded multi-chamber breech blocks were available, counted for nothing if the *Mitrailleuses* were used as light artillery. On the rare occasions when these guns served in an infantry-support role during the Franco-Prussian War, their sustained rate of fire of 100rd/min extracted such a terrible toll that the Prussians nicknamed them *Hollenkanone* – the 'hell-gun' – which, because the bores were parallel to the central axis instead of divergent, concentrated all 25 shots in such a way that men could be literally cut in half at short range.

Benjamin Hotchkiss, who had been granted a patent for a conversion of the Chassepot needle-rifle to fire metal-case ammunition, visited Paris in 1867 to attend the Exposition Universelle. Seeing commercial potential, he set up in business in 1867 in the commune of Viviez, north-west of Rodez in the Aveyron Département of southern France (Buigné & Jarlier 2001: I, 220), eventually to supply cartridges for Remington, Springfield-Allin and other US-made guns which were to be pressed into French service.

The enterprise proved to be so successful that, when the Franco-Prussian War ended in May 1871 with the capitulation of France and the formation of the Deutsches Reich, Hotchkiss elected to remain in France. His manufactory in Viviez had soon been moved to Saint-Denis, then an outlying district of Paris, where Établissements Hotchkiss & Cie were to make the American inventor's fortune. The Fusil d'Infanterie Modèle 1874, a conversion of the Chassepot needle-rifle to take metallic cartridges and better known as the Gras in honour of General Basile Gras, the Commandant of the École Normale de Tir de Châlons who (as a colonel) had been responsible for its development, was clearly inspired by the Hotchkiss patents, and the 11×59mmR Gras service cartridge owed much to the work that had been undertaken first in Viviez and then in Saint-Denis.

DEVELOPMENT
From revolver cannon to machine gun

ORIGINS

Benjamin Hotchkiss developed and then patented a handcrank-operated cannon in addition to his work on Chassepot needle-rifle conversions. The cannon looked similar to the Gatling Gun but was very different in detail; though both guns relied on rotating barrel clusters, the Hotchkiss had only a single bolt – simpler and more robust – instead of one for each barrel. Chambered for a 37mm self-contained cartridge, production-model Hotchkiss revolver cannons generally had five barrels and were potentially devastating.

Many French ordnance officers could still see the value that lay in a machine gun, and the French Navy, in particular, became increasingly keen to find an answer to the growing threat posed by the torpedo boat. The *Declaration Renouncing the Use, in Time of War, of Explosive*

Hotchkiss single-barrel cannon.
(Morphy Auctions,
www.morphyauctions.com)

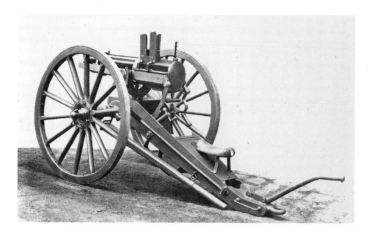

A typical five-barrel 37mm Hotchkiss crank-operated revolver cannon on its field carriage. This image was published in the British *Engineering* magazine in September 1892. (Author's Collection)

Projectiles Under 400 Grammes Weight, as proclaimed by the St Petersburg Conference of November–December 1868, had prohibited the use of shells weighing less than 14oz (397g) against human targets because to do so was considered to be inhumane, so Hotchkiss developed a 16oz (454g) shell to chamber in his 37mm ('1 Pounder') Revolver Cannon.

By 1890, Hotchkiss revolver cannons were being made for naval use (37mm, 47mm and 53mm) and, for land service, as a 37mm field gun or as a 40mm flank-defence gun with barrels rifled differently so as to enhance shot-dispersion. The cannons were all five-barrelled, and could be fed continuously simply by dropping cartridges into the hopper projecting diagonally upward from the breech. A typical naval-service 37mm Hotchkiss Revolver Cannon was 1,485mm long and weighed about 350kg on its shielded pedestal mount. The standard shell weighed 454g; muzzle velocity was approximately 410m/sec. Rate of fire was normally approximately 60rd/min, but could be raised to 90rd/min for short periods. Wheeled carriages were provided either to allow shipboard guns to be dismounted or armies to take them into the field. Once again, however, the military authorities rarely grasped the significance of these ideas and, therefore, saw no use for the Hotchkiss and comparable large-calibre weapons other than as light artillery.

While Hotchkiss laboured to perfect his revolving-barrel cannon, the first steps were being taken to harness the power generated within a gun as it fired. The most obvious way was to use the backward movement that could be seen in any free-mounted gun to operate its mechanism: opening

This British Maxim Gun, on its wheeled carriage, shows not only how the early machine guns were regarded as light artillery, but also their resulting vulnerability in battle. Note how the firer has had to crane his neck to use the sights. (Author's Collection)

BENJAMIN HOTCHKISS

Benjamin Berkeley Hotchkiss was among the giants of 19th-century weapons technology, founding what was to become one of the largest and most important European arms-making businesses and responsible for a series of tube-magazine bolt-action rifles and an efficient mechanically operated multi-barrel cannon.

Though the Hotchkiss family is now often claimed to have descended from French Huguenot refugees, Samuel Hotchkiss, who landed in North America in 1628, was English, having been born near Whitchurch in Shropshire. Benjamin Hotchkiss was born on 1 October 1826 in Litchfield, Connecticut, to metalsmith Asahel Augustus Hotchkiss and Althea Guernsey. The third of nine children, Benjamin, known to his friends and family as 'Berk', married schoolteacher Maria Harriet Bissell on 28 May 1850.

A.A. Hotchkiss & Sons had been making tools, agricultural machinery and ironware for many years in Sharon, Connecticut, and Andrew Hotchkiss (Benjamin's older brother) had even obtained a US Patent in 1855 to protect the well-known 3in (76.2mm) Hotchkiss Shell of the American Civil War. (The shell's two-part construction allowed what would now be called boat-tail construction, with a lead ring at the joint to seal the bore.) However, Andrew Hotchkiss died in 1858 aged just 37; Benjamin subsequently perfected the Hotchkiss Shell design by cutting grooves through the lead collar to allow propellant flash to ignite the paper-bodied time fuse, but more important internationally was the Revolver Cannon of 1872 that was fitted to many warships as an anti-personnel or anti-torpedo boat weapon.

Benjamin Berkeley Hotchkiss, 1826–85. (Author's Collection)

By the time of his death in Paris on 14 February 1885, Benjamin Hotchkiss had made a fortune estimated by his executors to be worth $12 million. Friends of his estranged wife claimed that he had bigamously married a 'Miss Cunningham' in Paris in 1867 and had also fathered a daughter who had died young. With his reputation being posthumously defamed almost as soon as he had been buried, it is difficult to accurately assess Hotchkiss's contributions to firearms history.

the breech, ejecting a spent case, reloading the chamber, re-locking the breech and then firing. It was also clear that this was potentially a continuous process.

To whom the honours are due is not so obvious, however. Recoil operation, usually claimed to be the older of the two principal systems, is credited to the American-born British inventor Hiram Stevens Maxim on the basis of his recoil-operated rifle with a toggle-lock breech mechanism that was protected by British Patent 8242/84 of 26 May 1884; the pioneers of gas operation are claimed to have been the Clair brothers (Benoit, Jean-Baptiste and Victor) of Saint-Étienne, whose French patent was granted in 1892. However, experimentation had been under way for many years. The American inventor Martin Pilon patented a self-cocking (though not self-loading) gun in Britain in December 1858 by harnessing recoil, and British Patent 1810/66 granted on 10 July 1866 to civil engineer William Curtis includes a clear description of gas operation.

ENTER ODKOLEK

Few of the books that feature 20th-century machine guns mention that the Hotchkiss, in its original form, was the work of Bohemian nobleman Adolf Odkolek Freiherr von Augezd. Still fewer have anything to say about the life and career of Odkolek, who is denied his rightful place in small-arms history.

Austro-Hungarian Privilegium 39/1378, sought on 22 February 1889, protected the bolt-action Neues Repetiergewehr, a repeating rifle designed by Odkolek and operated by sliding the trigger and trigger guard forward: a rack and pinion reversed the motion to open the bolt, which was then closed and locked by pulling the trigger guard back. A detachable box magazine hung on the right side of the feedway.

The Neues Repetiergewehr was too quirky to succeed. However, on 4 July 1889, Odkolek, through patent agent Heinrich Palm of Michalecki & Co., sought to protect a gas-operated machine rifle. The subject of Austro-Hungarian Privilegium 39/2887, this was probably the first truly successful gas-operated firearm as only the Clair-Éclair sporting gun seems to have been offered commercially by this time. Lack of success had been at least partly due to the novelty of smokeless propellant. Manufacturing problems led to fluctuating pressures and, therefore, inconsistent performance.

The prototype Odkolek gas-operated machine rifle was soon replaced by an improved version – Austro-Hungarian Privilegium 41/899 sought on 11 October 1890 and then further refined by Privilegium 43/313 of 24 March 1892 – but the basic operating principles remained largely unchanged. The Odkolek system was very simple. Gas was tapped from the bore, about two-thirds of the way to the muzzle, and allowed to strike a piston which formed part of the carrier. When the gun fired, gas pressure thrust the piston/carrier back, and cam surfaces on the carrier raised the locking struts (pivoted on the tail of the bolt) out of engagement with seats in the body walls. This allowed the bolt and carrier to run back to the limit of recoil; simultaneously, a cam-track in the piston extension acted with a pawl to move the next cartridge into place. As the return

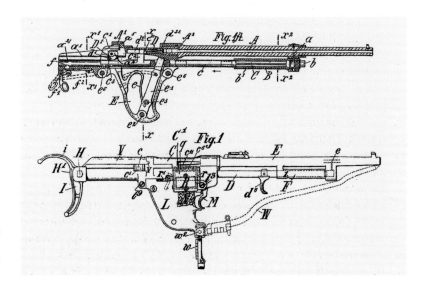

The gas-operated Odkolek light machine gun of 1891 (from US Patent 486938 of 29 November 1892) showing a feed strip and a bipod pivoted on the base of the trigger housing. (US Government Patent Office)

ADOLF ODKOLEK

Odkolek's family are mentioned in records dating back to 1383, and had settled in Obědovic und Ůgezdec in Bohemia by 1550. The title *Freiherr* had been conferred on Wilhelm Heinrich Odkolek von Ujezdec of Prague on 1 February 1680. (Král von Dobrá Voda 1904: 19). Adolf Odkolek himself was born into a Catholic land-owning family in Dub, near Strakonice, on 1 December 1854. On 30 September 1873 he joined the Austro-Hungarian Army's Ulan-Regiment 6 as an *Einjährige-Freiwilliger*, a one-year volunteer participating in a scheme intended to prepare potential officers for a military career. Having passed the cadet exam in November 1874, Odkolek was commissioned to command a cavalry detachment in May 1875, qualified for higher command in the officers' school (1877–78) and, after secondment to the Army shooting school, became regimental armourer officer in September 1881.

Odkolek's talents as a designer were soon evident, and, from 1 February 1882, he was encouraged to pursue his ideas, though often persevering without pay. Returning to service with Ulan-Regiment 3, Odkolek went on to serve as a squadron commander in Ulan-Regiment 11, and retired in 1896 ranking as *Rittmeister* (captain). Several inventions followed, including a tinder lighter, a semi-automatic breech mechanism for field guns, and the *Rationales-Gewehr* (Rational Rifle), patented in 1906 and offered to the Škoda company. When World War I began, Odkolek, by then aged 60, immediately volunteered for service. Initially appointed to command the reserve company of Dragoner-Regiment 11, he served as an instructor at the reserve officers' cavalry school in the summer of 1915 before being seconded to the Technische Militär-Komitee to develop grenade launchers and trench mortars.

Major d.R. ('of the Reserve') Adolf von Odkolek died suddenly on 2 January 1917. Buried in the family mausoleum in Friedhof St Helena, Baden bei Wien in eastern Austria, he was remembered in obituaries published on 9 February 1917 in *Reichspost* and *Neuen Freien Presse* as a writer and accomplished musician.

spring pushed the bolt and the piston/carrier unit forward, a new cartridge was pushed into the chamber.

HOTCHKISS TAKES CONTROL

The Odkolek machine rifle worked well enough to impress Hotchkiss & Cie, and it seems that rights, specifically to German Patent 65953, were purchased by Hotchkiss & Cie in 1893 or 1894 for next to nothing. Though Odkolek's crude prototype required considerable development to be considered battleworthy, the company was well aware of the effect the new automatic Maxim Gun could potentially have on profitability and was also actively seeking a replacement for the obsolescent manually operated Hotchkiss Revolver Cannon.

The Hotchkiss machine gun tested in 1895 was air-cooled at a time when most machine guns, such as the Maxim, relied on water cooling so that rapid fire, which heated even single-shot rifles until the chamber temperature rose high enough to ignite or 'cook-off' cartridges by heat alone, could be sustained. The idea of replacing a lightweight barrel when it got too hot, sometimes after only a few dozen rounds, quickly gave way to the use of a heavy barrel in which the heating effect, still present, reached critical levels much more slowly.

The perfected Hotchkiss also had radial fins on a bronze breech-sleeve to increase the surface area presented to the atmosphere. This undoubtedly improved the air-cooled weapon's ability to sustain fire, but water-cooled rivals still held an appreciable advantage if thousands of rounds were to be fired; most handbooks suggest that the Hotchkiss could fire only 10–12 cartridge strips before the barrel had to be cooled with water. However, it was far easier to manoeuvre and not as dependent on water supplies as the Vickers or Maxim guns.

Refining the Odkolek action was entrusted to Laurence Benét and Henri Mercié, both of whom were engineers employed by Hotchkiss. Benét was the son of Brigadier-General Stephen Benét, the US Army Chief of Ordnance for much of the life of the Trapdoor Springfield single-shot rifle. Laurence Benét trained as a mechanical engineer, and had gone to work for Hotchkiss & Cie in 1884 at the end of his graduation year. Benjamin Hotchkiss and Stephen Benét had enjoyed an amicable relationship, owing to contact made during the prolonged testing of the bolt-action Hotchkiss rifles, and it is assumed that this was enough to give Laurence Benét his opportunity.

Less is known about Henri Mercié, who rose to become Hotchkiss & Cie's chief development engineer. Working with Laurence Benét, he was responsible for perfecting not only the Odkolek machine rifle but also what is often known as the Fusil Mitrailleur Portative Hotchkiss, the 'Benét–Mercié Machine Rifle', one of the first machine guns light enough – more or less – to be handled if necessary by a single man.

Odkolek's feed mechanism was altered so that it could be driven from the reciprocating piston rod, and the design of the feed strips was greatly refined. Attempts had been made to develop cartridge belts, subject of at least two Austro-Hungarian patents, but the use of articulated metal strips was ultimately to be preferred. The strips allowed cartridges to be pushed forward into the chamber – unlike the Maxim Gun, which had to retract each cartridge and align it with the breech mechanically.

THE MODÈLES 1897 AND 1900

Introduced commercially in 1897 and offered in a variety of chamberings, the Hotchkiss was the first machine gun of its type to combine simplicity

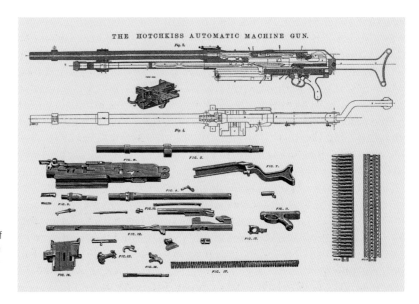

The 1897-pattern Hotchkiss machine gun. Note the position of the gas port and the length of the piston. This image was published in *Engineering* magazine in April 1897. (Author's Collection)

with reliability. Mounts ranged from simple *Affût trépied* (fixed-leg tripods), with provision for neither mechanical elevation nor traverse, to wheeled carriages with armoured shields.

M1900 Hotchkiss no. 2477 on tripod mount no. 2083. Note the shoulder stock attached to the rear of the body. (Morphy Auctions, www.morphyauctions.com)

The French Army purchased its first M1900 Hotchkiss machine guns in 1901, to arm two battalions of *chasseurs* (light infantry), followed by several hundred needed for extensive trials with the cavalry and garrisons in north-eastern France and North Africa. Many were still emplaced in fortifications or held in store at the outbreak of World War I in July 1914.

Manufactured in Saint-Denis by Hotchkiss & Cie, the Mitrailleuse Hotchkiss Modèle 1900 was similar to the M1897-type guns, but the gas system was improved. The gas tube had originally been full-length, drilled with radial gas-escape holes, but it was prone to clogging. The new gas tube was a half-length open-ended design with a 20-position gas regulator, which was so sensitive that a chart had to be issued with each gun to correlate ambient temperatures with regulator settings. The cocking handle was changed from a hook beneath the body to a handle on the left side, the barrel fins (usually five) were cast into a bronze collar around the barrel, and a radial safety lever was added to the left rear of the body. French-issue guns had tangent-leaf sights on top of the receiver above the bronze feed block, but others had a sight that was elevated by a knurled finger-wheel protruding obliquely behind the feed block.

The maker's plate on Hotchkiss no. 2477. The 'BHB' monogram was registered as a trademark on 12 May 1885, but had been replaced by crossed cannons by the early 1900s. (Morphy Auctions, www.morphyauctions.com)

According to data published in the British *Engineering* magazine, 20 February 1903, calibre could be 'any between 6mm and 8mm' and chambering could handle rimmed, semi-rim and rimless cartridges alike. Muzzle velocity varied from 600–750m/sec depending on the type of ammunition used. Metallic feed strips usually held 24 or 30 rounds. The guns featured in *Engineering* in 1903 included a regulator which could set the cyclic rate to 100rd/min (low) or 500–600rd/min (high).

THE M1900 EXPOSED

8×51mmR Modèle 1900 Hotchkiss

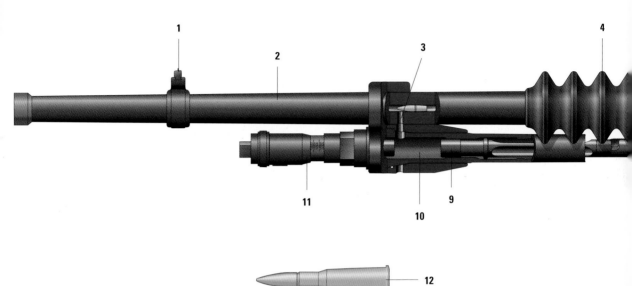

1. Front sight	**9.** Gas piston head	**17.** Rear sight
2. Barrel	**10.** Gas cylinder	**18.** Locking flap
3. Gas port	**11.** Regulator assembly	**19.** Gas piston and piston extension
4. Radiator cooling fins	**12.** 8×51mmR cartridge (not to scale)	**20.** Firing pin actuating lug
5. Breech cover	**13.** Feedway	**21.** Recoil (or return) spring
6. Shoulder piece or butt	**14.** Case deflector	**22.** Gun body
7. Pistol grip	**15.** Breech block	
8. Trigger	**16.** Firing pin	

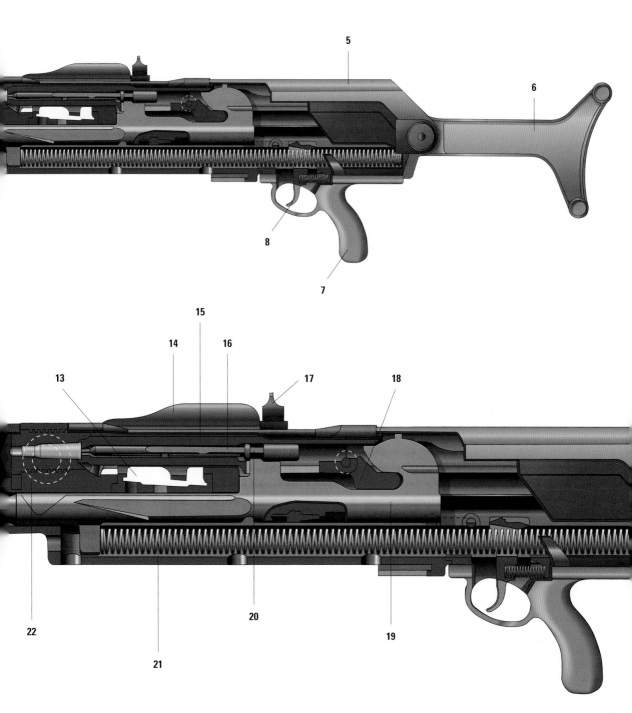

HOW DOES THE HOTCHKISS WORK?

All Hotchkiss machine guns are gas operated, relying on a long-stroke piston rod to operate the breech mechanism, though the locking systems of the M1900/M1914 and the Portatives differ. Starting with an empty gun, the mechanism is cocked either by pulling back on the cocking lever (located under the front part of the body of the M1900, but on the left side on the M1914) or the bolt-handle of the Portative and Benét-Mercié guns. The bolt can be set for single shots or automatic fire, but this affects only the trigger mechanism.

A stamped-sheet *bande rigide* (feed strip), with three parallel rows of clips to hold the cartridges in place, can be inserted in the feedway and pushed firmly until it locks in place. When the trigger is pressed, the breech closes. A cartridge is pushed forward into the chamber, the locking system is activated, and the firing pin is released to strike the primer of the chambered cartridge.

A tiny portion of the propellant gas following the bullet, tapped from the bore about half way to the muzzle, strikes the piston head. This forces the piston unit backward against the recoil spring. The backward movement cams the locking flaps out of engagement with the body (M1900/M1914) or revolves the rotating locking collar ('fermeture nut') to release the breech block (Portative). The spent cartridge case is ejected as the breech opens, and a new cartridge is moved into place by the feed pawls. Unless held back by the sear, the recoil spring then reasserts itself to close the action, strip a new cartridge into the chamber, lock the breech and fire once again. The breech is held back during lulls in firing – the so-called 'open breech' system which helps to keep the chamber as cool as possible when firing automatically.

Gas operated, and locked by pivoting struts on the bolt engaging the receiver walls, the M1900 Hotchkiss was 1,360mm long, had an 825mm barrel, and weighed about 23kg without the tripod that added 17.9kg. There was also a large-calibre version, chambered for a 12mm (0.472in) cartridge developing a muzzle velocity of 616m/sec, which fed from 20-round strips. *Engineering* claims that this variant had been developed as a result of experience in the Second Anglo-Boer War (1899–1902), when the striking power of rifle-calibre guns had proved insufficient at long range.

THE FIRST SUCCESSES

Adolf Odkolek had been forgotten by the time the Hotchkiss machine gun was not only being purchased by the French Army but also touted commercially. The M1897- and M1900-type Hotchkiss guns were sold to Japan in 6.5×50mm; to Norway, Portugal and Sweden in 6.5×55mm; to Brazil, Chile, Mexico, Peru, Spain and Venezuela in 7×57mm; to Belgium, Bolivia and Turkey in 7.65×53mm; to China in 8×57mm, and to New Zealand in 0.303in.

Among the most interesting was the Mitraljøse m/1898, now almost unknown outside Norway. After testing a Maxim Gun early in 1896 and a Hotchkiss in the late summer of 1897, the Norwegian machine-gun commission reported that the Hotchkiss, being both simpler and cheaper, was worthy of adoption. A prototype chambered for the 10.15×61mmR cartridge was ordered from France in January 1898 and a manufacturing licence was acquired from Hotchkiss & Cie. The first French-made gun was followed by another 26 for issue to the Norwegian fortress and coastal artillery and possibly two others for proof-firing trials. The first five 10.15mm m/1898 machine guns made in the state-owned Kongsberg

Våpenfabrikk had been delivered by October 1899, numbered 30–34, but the decision was then made to substitute the 6.5×55mm rifle cartridge so that the m/1898 could be issued more widely.

A 6.5mm-calibre test gun arrived from France in April 1899; comparatively minor changes were made after trials with the test gun proved encouraging, and the first series-made 6.5×55mm m/1898 was delivered from Kongsberg Våpenfabrikk on 5 December 1899. Production continued until 1925, by which time 170 m/1898s had been made (numbered 50–219). The m/1898 was 1,440mm long, had a 760mm barrel, and weighed about 25.2kg. Muzzle velocity was 740m/sec with the standard 6.5mm ball cartridge, and cyclic rate averaged 550rd/min. The 10.15mm m/1898 fed from a 25-round rigid metal strip, but the 6.5mm version could accept either a conventional 30-round metal strip or an articulated belt designed by a Norwegian armourer named Mørchs, which could hold 100 rounds. Distinctive marks include the crowned 'H7' cypher of the king, Haakon VII, and the 'HF' marks of arms-inspector Haakon Finne.

The 6.5×55mm m/1898 was superseded by the m/29, a a Kongsberg-made variant of the Colt MG38B (the commercial version of the M1917 Colt–Browning), but Hotchkiss machine guns – including a few 10.15mm-calibre examples – were still either in Norwegian second-line service or languishing in store when German forces invaded Norway in April 1940. Some of these weapons were undoubtedly, if briefly, fired in anger.

THE HOTCHKISS AND JAPAN

Hotchkiss machine guns served the Imperial Japanese Army (IJA) for many years. The first to be issued to the IJA were M1897s bought from Hotchkiss & Cie and possibly issued as the 'Meiji 30th Year Type' (though usually known as *Hō-Shiki Kikanjū*), but licensed production subsequently

Meiji 38th Year Type Hotchkiss no. 2048, made in Koishikawa in March 1913, mounted on a later Type 92 tripod with pole-sockets on each foot. Note that there are seven large barrel-cooling rings. (Morphy Auctions, www.morphyauctions.com)

Another view of Meiji 38th Year Type Hotchkiss no. 2048. (Morphy Auctions, www.morphyauctions. com)

began in Japan. The Hō-Shiki was readily identified by seven cooling rings on the barrel instead of the customary five.

The IJA's Hotchkiss machine guns were widely used during the Russo-Japanese War of 1904–05: for example, the three divisions involved in the attack on and subsequent siege of the Russian deep-water naval base at Port Arthur (August 1904–January 1905) each had 24 on strength. Though intended for use in defensive positions, the machine guns proved particularly valuable for offensive action, towed on sleds and manoeuvred over obstacles by hand. The Japanese learned the value of a machine gun for overhead and enfilade fire long before most Western armies came to similar conclusions in the trenches of World War I.

War suggested improvements that could be made, resulting in the Meiji 38th Year Type machine gun or M1905 made by Koishikawa Zoheishō in Tōkyō. Many features of the Hō-Shiki were retained, but the design of the tripod was noticeably different. The 38th Year Type was replaced by the Taishō 3rd Year Type of 1914, which, like its predecessors, chambered the 6.5×50mm Meiji 30th Year Type cartridge shared with the Arisaka infantry rifles and carbines.

Oddly, a variant of the Taishō 3rd Year Type made in Koishikawa Zoheishō, but with a Hotchkiss-supplied barrel chambered for the 7×57mm cartridge, was sold to Chile shortly after World War I had ended. At least 200 examples of the Ametralladores Modelo 1920 were acquired, apparently from Taihei Kumiai, a government-sponsored export agency trading in Tōkyō. The Chilean national arms will be found on top of the body.

FRENCH APATHY

The French military authorities, reluctant to pay Hotchkiss & Cie more than was absolutely necessary, elected to improve the M1900 machine gun. Though about 50 Hotchkiss guns were acquired in 1906, trials suggested, at least to the French Army, that the M1905 machine gun, credited to government technicians in the Ateliers de Puteaux and consequently known as the 'Puteaux' or APX, was preferable. The M1905 could be distinguished by the squared contours of the breech, similar to

AMMUNITION

The Hotchkiss was originally developed to fire the French 8×51mmR Modèle 1886 M cartridge, soon to be replaced by the Modèle 1886 à Balle D loaded with Captain Georges Raymond Desaleux's innovative boat-tail bullet which improved both range and accuracy. Both ammunition types performed well in the M1900-type Hotchkiss, but experience showed that the Balle D and the M1907 Saint-Étienne machine gun were poorly matched. A rise in the number of ignition failures was ascribed to primers being forced from case heads, so the Balle D a.m. (*amorçage modifié*, 'altered priming'), with more effectual crimping, was approved in 1912. However, the M1914 Hotchkiss also suffered jams until sufficient quantities of the Balle D a.m. became available towards the end of 1915.

There were Balle T (*traceuse*: tracer) and Balle P (*perforante*: armour-piercing) cartridges, loaded respectively with an 11.2g bronze bullet containing trace compound in a central cavity and a 9.6g steel-cored bronze bullet. Muzzle velocity, 750m/sec and 840m/sec respectively, was appreciably greater than the 701m/sec of the 12.8g Balle D, which compromised the utility of mixed-ammunition strips by considerably altering bullet trajectories. There was also a special blank cartridge, Balle 1905 B (*blanc*: blank), loaded with a wood bullet and a greatly reduced propellant charge, for use with the *bouchon de tir à blanc* (BTB: blank-firing attachment).

Twelve 24-round ammunition feed strips were carried in the *caisse en bois*, a wooden case with sheet-metal protective edging, divided internally into six two-strip compartments. The lid was usually hinged on the long edge (short-edge examples are known) and closed by a push-button spring latch. Boxes usually display '12 BANDES', '288 CARTOUCHES' and a large 'H' for Hotchkiss stencilled on the front face.

A similar box was made for the *bande articulée*, the articulated belt developed during World War I for use in the air and then pressed into ground service as a way of increasing the Hotchkiss machine gun's ability to sustain fire. Similar in construction to the 12-strip case, the box had a combination catch/lever with a clutch which could be engaged to wind the belt around a central spindle or disengaged to allow the belt to be pulled freely out of the box. The belt was made up of 83 three-round striplets, though the cartridge nearest the box spindle was usually omitted so that the first striplet formed the base of the triangle around which the remaining striplets could wind. Capacity, though usually noted as 250, was in practice only 248 rounds.

Ammunition carried by each French Army *section de mitrailleuse* (machine-gun section) amounted to 41 cases (11,466 rounds) for immediate use, and 762 *bandes rigides* (18,288 rounds) in the *caisson de ravitaillement* or limber. After the *bandes articulées* had been introduced, nine 'ready use' boxes containing them were substituted for nine of the rigid-strip type.

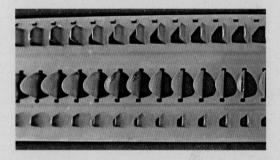

The M1914 *bande rigide*, showing the rows of retaining clips. (Morphy Auctions, www.morphyauctions.com)

those of the later M1907 Saint-Étienne machine gun. Small-diameter fins ran half the length of the barrel, and the return spring, concentric with the barrel, ran from the front-sight block to a collar at the muzzle. The gas port was moved to the muzzle, and a new adjuster was theoretically capable of altering the rate of fire from 8rd/min to 650rd/min.

The changes incorporated in the APX were expected to improve cooling, but the fins proved to be little other than a trap for oil and dirt. When any prolonged firing was undertaken, not only did the barrel overheat but a haze of oil vapour disturbed the sight picture. In addition, though the APX retained Hotchkiss-style feed strips, these were not interchangeable with the original form – particularly unfortunate, as M1900-type Hotchkiss guns were still serving with French Army cavalry units and in the French colonies. The APX mount was either the M1900-type Hotchkiss tripod or the M1905 quadrupod, the latter with distinctively curved legs.

French soldiers with an M1905 Puteaux machine gun mounted on an M1915 Type Omnibus tripod. Note the plain-sided body, the finned barrel and the return spring coiled ahead of the front-sight collar. Note also the empty feed strip by the gunner's foot and the canvas bucket. (Author's Collection)

The troublesome Puteaux machine guns were rapidly withdrawn from the French field army and relegated to static use in fortresses and strongpoints, where their deficiencies were not as obvious. They were superseded by the M1907, credited to technicians in the Saint-Étienne ordnance factory, which began to reach the troops in 1909.

The change to the M1907 Saint-Étienne had been partly due to the publication in 1907 of *Sommes-nous defendus?* ('Are we defended?'). The author, Charles Humbert, a newspaper proprietor who had served as a captain in the French Army, criticized the employment of machine guns in the French Army by drawing attention to German superiority. There was such a surge of popular opinion in support of Humbert that the Assemblée nationale became involved and, on 27 September 1907, the members of the conseil supérieur de la guerre approved the creation of

M1907 Saint-Étienne no. 11223 of 1916, shown here on an M1915 Type Omnibus tripod and fitted with the original spatulate type of flash hider, was an adaptation of the original Hotchkiss. The streamlined slab-sided body and plain barrel are obvious distinguishing features. (Morphy Auctions, www.morphyauctions.com)

An M1907 Saint-Étienne gun crew from the 124e régiment d'infanterie. The photograph was clearly taken some time after World War I had ended, as a fabric ammunition belt is being used. (Author's Collection)

sections de mitrailleuses (machine-gun sections) attached to regiments and brigades.

French military tactics were swinging back from largely static to *guerre à l'outrance* ('all-out war'), and an air-cooled machine gun promised an acceptable compromise of firepower and mobility. General Henri de Lacroix, vice-president of the conseil supérieur de la guerre in 1907–09 and the promoter of the French mobilization plan of 1909, opined that the ready availability of machine guns would improve morale.

Inspiration was provided by the Hotchkiss machine gun, but changes were made in a quest for a more battleworthy weapon. Not all of these were advantageous, and the M1907 Saint-Étienne has had more than its fair share of critics. Though potentially an improvement on the M1900 Hotchkiss in several respects (most notably the ease with which the barrel could be changed), the toggle-joint lock and the operating mechanism were grave errors. The goal may have been reduced recoil-effect, but this forced the designers to introduce a rack, pinion and cam-plate mechanism to open the breech. The return spring, inside the lower rear of the body in the original Hotchkiss, was now exposed to heat and dirt beneath the Saint-Étienne barrel.

The metal-strip feed was also unwise, particularly as it could not be interchanged with the Hotchkiss design (the Saint-Étienne withdrew cartridges backwards before raising them in line with the chamber), and the incorporation of rate regulators showed that little had been learned from the design shortcomings of the M1905 Puteaux. The Saint-Étienne not only had a lever on the rear left side of the breech to select high or low rates of fire, but also a regulator to vary the rate within each of these settings. This theoretically allowed a cyclic rate of 10–600rd/min.

The M1907 Saint-Étienne had a streamlined appearance, with a slender barrel and a single handgrip attached to the back plate of the body. A tangent sight, inspired by the German Lange design, lay on top of the breech. The standard mount was either the Affût-trépied 1907 C or the Modèle 1915 Type Omnibus. However, a special forked pillar was made for trench service; when used in this way, the Saint-Étienne was fitted with a butt extension instead of a spade grip. Combat experience revealed design and operational shortcomings and an improved M1907/16 Saint-Étienne appeared.

THE HOTCHKISS PORTATIVE

Virtually every military power, France included, had sent observers to the Russo-Japanese War in the hope that valuable lessons could be learned from a military clash between two powerful states. The value of the machine gun became obvious almost as soon as fighting commenced: static emplacements that favoured the Imperial Russian Army's Maxims and, to a perhaps lesser extent, the IJA's Hotchkisses, and the use of the Danish Madsen light machine gun – virtually a large automatic rifle – by Russian cavalry and other specialized units as the war ran its course.

Laurence Benét and Henri Mercié began work on a lightweight machine gun that could compete with the Madsen. The outcome was the Fusil Mitrailleur Hotchkiss, Modèle 1908, often known simply as the 'Hotchkiss Portative'. Attempts to lighten the original Odkolek-inspired Hotchkiss machine gun failed to impress, and so the new design relied on a rotating locking collar ('fermeture nut'), containing an interrupted screw, which engaged threads in the body to lock the action. Experiments proved this locking mechanism to be effective, and protection was sought for the new design: an application for what was to become French Patent 377717 was made on 13 September 1907, and comparable requests were submitted in Britain, the United States and elsewhere. Unlike some patents, which condensed everything into a few pages, Benét and Mercié's specifications were minutely detailed. For example, US Patent 861939 of 30 July 1907 runs to 43 pages.

The feed system of the original Hotchkiss was perpetuated, but the feed strip was inverted so that the cartridges were protected from rain. However, loading was unquestionably more difficult; so awkward, indeed, that claims were subsequently made that firing was only possible in daylight.

The patent drawings show a lightweight plain-surfaced barrel protected by a ventilated sheet-metal handguard doubling as a fore-end. Testing revealed that the barrel overheated all too easily, and that the handguard not only got too hot to hold but was also susceptible to damage from blows. A heavy partly finned barrel was soon substituted, the idea of a forward handgrip being abandoned.

Chambered for the standard 8×51mmR rifle cartridge, the gas-operated Fusil Mitrailleur Modèle 1908, adopted by the French Army as a cavalry weapon, fed from a 30-round rigid *Bande chargeur* (feeder belt).

The gun was 1,190mm long and weighed 12.5kg with its bipod. The barrel was 565mm long. A monopod could be fitted beneath the butt if required. Muzzle velocity was 650m/sec with standard ball ammunition, and cyclic rate averaged 500rd/min. The original tangent-leaf rear sight was graduated to 2,000m.

The M1908 Portative was greeted with enthusiasm, as it was lighter and far easier to handle than the regulation M1907 Saint-Étienne though still heavier than the Madsen. Changes were apparently made to the sights and the bipod in 1913, creating the so-called 'M1908/13', but it has been suggested that only 110 of these had been acquired by the French Army by the summer of 1914 (Scarlata 2019, quoting Jean Huon), to allow extensive trials to be undertaken, though perhaps 500 more were pressed into service when World War I hostilities began.

A few Portatives saw service in the air, some were fitted to armoured cars, and others survived the battles that raged around Verdun in 1916. However, most of them were soon replaced by the British Lewis Gun or the infamous Chauchat light machine gun. Production of the Portative had ceased by 1915, as the French government was committed to developing other designs – particularly the execrable Chauchat – that were thought to be better suited to the French tactic of 'assault-at-the-walk'. Consequently, Hotchkiss & Cie may have been allowed to sell the remaining guns and components to Britain. As the Hotchkiss was much easier to make than the Vickers or Maxim guns, production began in Britain in 1916.

Portatives made by Hotchkiss & Cie had also been sold prior to 1914 to Belgium (in 7.65×53mm), to Brazil, Chile and possibly other armies in South and Central America (7×57mm), and to Sweden (6.5×55mm). Judging by production serial numbers, Brazil and Chile may each have acquired at least 1,000 Portatives.

The Hotchkiss Portative was also made in Norway alongside the Mitraljøse m/1898. A test gun had been ordered from Hotchkiss & Cie in February 1908, when precise details of the Norwegian cartridge and chamber dimensions had been sent to the Saint-Denis manufactory,

Brazilian 7×57mm M1908 Portative no. 144. It seems that as many as 1,000 of these light machine guns were purchased by Brazil shortly before World War I began. Note the position of the bipod and the design of the butt. (Morphy Auctions, www.morphyauctions.com)

The Brazilian national arms found on Portative no. 144. (Morphy Auctions, www.morphyauctions.com)

allowing trials to be undertaken in October that year in the presence of Laurence Benét and one of his technicians. Problems included gas leaks from the locking collar, but a modified machine rifle delivered in June 1909 proved good enough to allow both official adoption and acquisition of a manufacturing licence.

Orders for 25 6.5mm Let Mitraljøsen m/1911 for the Norwegian Army were placed in February 1911 and December 1913, but these, and a further four ordered for the Norwegian Navy's torpedo boats, were never entirely successful. Numbered 1–54, they were supplanted by the Madsen m/1914, though a few that survived in 1940 served the German invaders as 6.5mm Leichte Maschinengewehre 101(n). The m/1911 could be identified by marks including '6.5$^{m}/_{m}$ HOTCHKISS' above 'LET MITR' on the left side of the body, by the crowned 'H7' royal cypher, and arms-inspector Haakon Finne's 'HF' inspection marks.

THE BENÉT-MERCIÉ MACHINE RIFLE

Many US Army officers recognized the value of the .30-calibre Gatling Guns directed by Lieutenant John Parker at the Battle of San Juan Hill (1 July 1898) and the M1895-type Colt–Browning 'Potato Digger' machine guns issued to the US Navy and Marine Corps during the Spanish–American War of 1898 (the US Army had about 200 examples of the M1895 on strength, but none saw active service during this conflict). The most progressive thinkers determined to introduce something that was less clumsy than the .30-calibre M1904 Maxim, and tests were undertaken in 1908 involving the Maxim, an M1908 Benét-Mercié of the type that had been approved by the French Army as a cavalry weapon, and the Pratt & Whitney 1902-pattern De Knight – an interesting but overly complicated gas-operated weapon designed by Victor De Knight of Washington, DC. Constant jamming and too many broken parts caused the De Knight to be rejected, leaving only the Maxim, which had apparently been excluded from the first series of trials on the basis that it was too clumsy, and the Benét-Mercié.

US Army M1909 Benét-Mercié Machine Rifle no. 576. The 'US Machine Rifle, Caliber 0.30-inch, Model of 1909' was 46.9in (1,191mm) long, had a 22.2in (564mm) barrel with four-groove left-hand twist, and weighed 27.7lb (12.6kg) with its bipod. A 30-round metal feed strip was used; muzzle velocity with the M1906 cartridge was approximately 2,800ft/sec (853m/sec); cyclic rate averaged 500rd/min. The M1909 had a bipod, a combined shoulder-stock/pistol grip and a sophisticated butt-monopod with a crossbar carrying two small pads. This gave a surprisingly stable firing position, even though the bipod was flimsy and often stayed by a leather strap passed through the trigger guard to loop around both legs. The monopod was the work neither of Springfield Armory nor US Army technicians, but was patented by Hotchkiss & Cie (INPI 1908). (Morphy Auctions, www.morphyauctions.com)

M1909 Benét-Mercié Machine Rifle no. 368, with a Warner & Swasey M1913 5.2× sight, is mounted on the rarely encountered and largely experimental Mark II tripod. Two types of tripod were developed for the gun, but neither reached service status and production was meagre. (Photo courtesy of Cowan's Auctions Inc., Cincinnati, OH)

Eventually, after both the Hotchkiss and Benét-Mercié machine guns had proved to be acceptably efficient, a trial was undertaken under the supervision of Major George W. McIver, commandant of the Army School of Musketry, in which 110 life-size figures – prone, kneeling and standing – were to be engaged at a range of about 1,000yd (914m). The Benét-Mercié achieved 43.8 per cent hits compared with 40.6 per cent for the M1904 Maxim, and, owing to its portability, was declared victorious. The US Army Chief of Ordnance, Brigadier General William Crozier, concurred and the Benét-Mercié Machine Rifle, a comparatively minor modification of the French M1908 for the .30-06 cartridge, was adopted officially in 1909. While production was being considered, 29 M1908 Portatives, presumably adapted for the .30-06 cartridge, had been acquired from France so that tactics and training methods could be developed.

It has been widely reported that 1,070 guns were made in Springfield Armory with the assistance of Colt: 670 for the US Army, 400 for the US Navy and the Marine Corps. However, evidence provided by US Ordnance Corps records and analysis of serial numbers is conflicting. There is currently no indication that the Benét-Mercié Machine Rifles were numbered in one sequence, or if the US Navy/Marine Corps guns, which were designated 'MARK II' (with 'MOD. 1' added after upgrading) were considered separately.

Opinion was divided as to the merits of the Benét-Mercié Machine Rifle. Lieutenant John 'Gatling Gun' Parker declared it to be the most perfect mechanism of any machine gun yet invented, but Douglas MacArthur, at that time a captain in the General Staff Corps, took the view that the performance of the new machine gun did not match its theoretical capabilities. Experience suggested that MacArthur's view was closer to the truth, at least until some of the most obvious problems – which included persistent extractor and ejector failures – had been overcome.

The overly complicated rear sight of the M1909 Benét-Mercié Machine Rifle reflected the view that accurate fire was obligatory even in a machine gun. Note the multi-aperture disc and the in-built correction for bullet drift. (Morphy Auctions, www.morphyauctions.com)

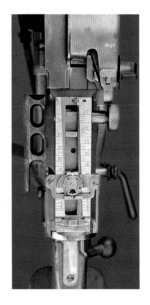

25

.30-06 Benét-Mercié Machine Rifle M1909

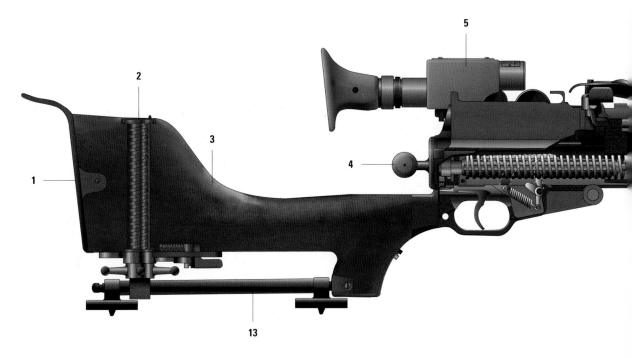

1. Buttplate
2. Monopod elevator
3. Buttstock
4. Cocking handle
5. Warner & Swasey optical sight
6. Radiator fins
7. Gas port
8. Barrel

9. Front sight
10. Bipod
11. Regulator assembly
12. Piston head
13. Monopod bar
14. .30-06 cartridge (not to scale)
15. Feed cover and deflector plate
16. Fermeture nut

17. Firing pin in breech block
18. Handguard
19. Actuator
20. Cocking handle shank
21. Sear
22. Trigger

This US Navy Benét-Mercié Machine Rifle, Colt-made no. 537, shows not only that it is an improved pattern ('MARK II') but also that it has been upgraded: 'MOD. 1' quite clearly being an addition. The 1912-dated inspector's mark was applied by Lieutenant Nelson H. Goss. (Morphy Auctions, www.morphyauctions.com)

The 1910 *Annual Report of the Chief of Ordnance* noted that the manufacturing specifications for the M1909 Benét-Mercié Machine Rifle had been prepared, while the 1911 report claimed that delivery of the first 100 guns was to be completed by 1 October that year. A locking device for the elevating mechanism of the monopod and a blank-firing attachment had also been approved. About 200 guns were issued to the 'Mobile Army' in 1912, replacing M1904 Maxims, and a manufactory capable of making 1,000 feed strips daily had been created. Long-term endurance tests, expending more than 200,000 rounds, had been completed with almost no problems and, in 1913, 335 Warner & Swasey 5.2× M1913 optical sights were fitted.

In 1914, not only were 615 M1909 Benét-Mercié Machine Rifles overhauled and upgraded – 245 of them from the US Navy and Marine Corps – but two types of Maxim silencer had been subjected to exhaustive tests. The ten-chamber Type A silencer was preferred, as it suppressed the sound of firing more effectively, but nothing more was done.

JAPANESE DEVELOPMENTS

Just as World War I began, the Japanese authorities realized that the action of the Taishō 3rd Year Type Hotchkiss, which had replaced the Hō-Shiki and the Meiji 38th Year Type machine guns, was unreliable with the full-charge load. Instead of revising the gun to retain standard ammunition – which would have been advisable – they decided to retain the gun and simply reduce the power of the cartridge to delay the speed at which the breech opened.

The Japanese 6.5×50mmSR cartridge was intended for a manually operated rifle in which the chamber pressure dropped to virtually nothing before the breech was opened, but the machine gun opened so quickly that residual pressure within the cartridge case tended to hold it against the chamber wall. If this happened, the extractor could tear through the case rim or even separate the case head from the neck. The 3rd Year Type Hotchkiss also sometimes suffered from excessive headspace, a manufacturing problem which allowed the base of the cartridge case to be set back against the bolt face when the breech opened.

The advent of the special '3rd Year Type' cartridge was an unwanted complication. Lubrication was still needed to ensure that the parallel-sided cartridge cases could slide more easily into the chamber and then extract satisfactorily. The addition of an oiling pad, which coated each cartridge before it entered the breech, improved extraction without resorting to pre-lubricated ammunition, but still attracted grit

The original Hotchkiss-type ejector was replaced by a Lewis pattern, which threw the spent cases out over the bolt. Otherwise, the 3rd Year Type action was much like that of the Meiji 38th Year Type, with similar gas operation and a displaced bolt-flap lock. Spade grips replaced the pistol grip, however, and the barrel was enveloped by large cooling fins.

USE
The Hotchkiss at war

The use of the machine gun in combat can be traced back to Gatling Guns and others used during the American Civil War, to the employment of the de Reffye Mitrailleuse during the Franco-Prussian War, and to the widespread issue of manually operated Gatling, Gardner and Nordenfelt guns for land and maritime use in the 1870s and 1880s.

The British Army had used Maxims in Matabeleland and Mashonaland during the First Matabele War of 1893–94, in British India during the Chitral campaign of 1895, in Sudan during the Second Sudan War of 1898 – to good effect during the battles of Atbara and Omdurman – and then in South Africa during the Second Anglo-Boer War of 1899–1902. However, it is arguable where the Hotchkiss machine gun first saw combat.

JAPAN AND CHINA

When war with China began in 1894, with the Japanese attempting to free Korea from Chinese domination (hoping to replace it with Japanese primacy), both sides were short of modern weapons. The Japanese eventually decided to raise an expeditionary force to occupy the island of Taiwan, and so, on 6 March 1895, an infantry regiment – reinforced so that it mustered 2,800 men and an artillery battery – sailed from Ujina to Sasebo, and then, under naval escort, landed on the Pescadores Islands in the Taiwan Strait on 23 March. The official history of the campaign is said to have claimed that the expeditionary force had a battery of machine guns fighting alongside the field artillery.

Two hundred Maxims had been hurriedly made in Japan – whether with or without the benefit of a licence is unknown – by the artillery arsenals in Tōkyō and Ōsaka to arm eight six-gun batteries serving the

Imperial Guard and the 4th Division of the Japanese Second Army. The guns were mounted on wheeled carriages, and accompanied by lightweight limbers. They apparently chambered the 8mm Meiji 22nd Year Type cartridge, loaded with black powder which encouraged fouling, and so were judged to be ineffective in combat. The *Ma Shiki Kikanjū*, as the Maxim was known, was subsequently replaced by Hotchkiss guns chambering the 6.5×50mmSR smokeless Meiji 30th Year Type cartridge, though the older guns were still in service with the Imperial Guard in 1900.

If doubts remain concerning the participation of machine guns during the fighting on Taiwan and in Korea, there can be no doubt that the Japanese used them in China during the Boxer Rebellion (1899–1901). On 2 September 1900, *The Shanghai Mercury* newspaper reported:

> Amoy, 29th August. Japanese Landed. On Saturday, the 25th inst., at 1 p.m. the Japanese men-of war landed on the British concession 100 men fully armed, with one machine gun, and placed sentries at the Taiwan Bank ... The sailors were not withdrawn before Sunday morning, when they with others, 250 strong, with two machine guns, landed to the south of the British concession ...

The Japanese warships carried Maxims, mounted on pedestals and in fighting tops as anti-boarding weapons. However, the guns could usually be dismounted to be sent ashore with landing parties.

Claims have been made that at least one Hotchkiss was mounted on armoured trains used by the British during the Second Anglo-Boer War, particularly one built in Mafeking during the siege of that town, but photographs show that manually operated two- and five-barrel Nordenfelt multi-barrel guns were used alongside Maxims. The Hotchkiss may have been a single-barrel 37mm cannon.

THE HOTCHKISS IN THE RUSSO-JAPANESE WAR

The Russo-Japanese War of 1904–05 was the first in which 'automatic' machine guns were used in quantity by the armed forces of one major power against those of another. The war itself was basically territorial. The Russians had effectively annexed Manchuria, installing a huge army to protect the construction of the Russian-owned Chinese Eastern Railway, while Japan, seeking to exert influence over Korea, saw Russian encroachment as a threat. Beginning in February 1904, the war raged on until brought to an end in September 1905 by the Treaty of Portsmouth. Hundreds of thousands of men had been engaged, but crushing defeats inflicted on the Russians on land at Mukden (20 February–10 March 1905) and at sea in the Tsushima Strait (27–28 May 1905) had proved to be conclusive. The Japanese were victorious.

Among the many military observers who had travelled to witness how the Russians would overcome the Japanese, those who retained an open mind could see that the machine gun had come of age. British Army Captain James B. Jardine reported that:

The Hō-Shiki was a Japanese-made variant of the French-supplied M1900 Hotchkiss, readily distinguished by its seven cooling fins instead of the customary five. The simple tripod, with no provision for traverse, was retained. (Author's Collection)

All officers are enthusiastic about them. All agree that their chief rôle is defense, even at night, and that they are extremely useful in attack [the Japanese used a light, Hotchkiss-type machine gun that was much more portable]. During the battle of Mukden machine guns were used very much in the attack by the Japanese, but it seems that the casualties of the machine gun detachments were very heavy indeed; one commander thought them especially useful in pursuit. (War Office 1908: 346)

The Russians had purchased 379 Maxims in 1896, intending to undertake extensive trials, and obtained a manufacturing licence from Vickers, Sons & Maxim in 1902. Consequently, Russian Maxims were pitted against the Japanese *Hō-Shiki Kikanjū* M1900-type Hotchkisses from the commencement of hostilities.

The Russo-Japanese War (overleaf)

Russian Primorskiy Dragoons – readily identified by their yellow shoulder boards, with Cyrillic 'Prm' in red – attack a Japanese entrenchment near Port Arthur in the early winter of 1904. The riflemen carry 6.5mm Meiji 30th Year Type Arisaka rifles, with the 38cm-bladed bayonets fixed. They are wearing greatcoats to protect against the cold, with the waist belt carrying ammunition pouches and the bayonet frog worn outside the coat. A machine-gun crew has brought its Hō-Shiki Hotchkiss into action to provide fire support. The firer has a thin yellow ring above a medium-width gold ring around his cuff, showing his rank of corporal or *go-shō*. The dragoons wear regulation M1897 dark-green tunics and black fur shapka caps. They are wielding the M1881 dragoon sabre, but also carry M1891 Mosin-Nagant dragoon rifles over their shoulders, and ammunition boxes on their belts. The Russian officers, a captain or *Rotmistr* and a junior lieutenant, *Praporshchik*, carry M1895 Nagant revolvers holstered on their belts, secured with lanyards of silver thread and interwoven in orange and black, around their necks.

Eight Maxims, mounted on high, large-wheeled carriages, equipped a machine-gun battery which customarily accompanied the Russian field-artillery batteries. The Japanese Hotchkisses were issued to cavalry brigades, usually six guns per battery instead of eight, though some infantry units had machine-gun sections with two Hotchkisses apiece. Their tripod mounts, though still surprisingly heavy, permitted far better mobility than the Russian carriages. At the Battle of the Yalu River (30 April–1 May 1904), the annihilation of the Russian Maxims by Japanese field artillery had been attributed to the Russian guns' visibility. Wheeled mounts were promptly replaced by tripods.

The utility of the machine gun was gradually recognized by Russians and Japanese alike. Fighting Cossacks, who were physically bigger, rode more powerful horses and used the lance most effectively, soon showed Japanese cavalrymen the futility of engaging in traditional horseman-against-horseman combat. A better solution was to dismount, bring up machine guns, and confront Russian cavalrymen as infantrymen. The Battle of Wa-Fan-Gou (14–15 June 1904) showed the devastating effects that a few well-sited and boldly handled machine guns could have:

> The machine guns were extraordinarily successful. In the defence of entrenchments especially they had a most telling effect on the assailants at the moment of the assault. But they were also of service to the attack, being extremely useful in sweeping the crest of the defenders' parapets; as a few men can advance under cover with these weapons during an engagement, it is possible to bring them up without much loss to a decisive point. The fire of six machine guns is equal to that of a battalion [*sic*] ... (Longstaff & Atteridge 1917: 47)

The Battle of Sha-Hō (5–17 October 1904) is generally regarded as the turning point of the Russo-Japanese War, particularly in the Imperial Japanese Army.

> Stealthily manoeuvring his six machine guns into position on a high and broken spur which ran down to the water's edge, he [Prince Kanin Kotohito, commander of the Japanese 2nd Cavalry Brigade] suddenly opened a hellish rain of bullets upon two Russian battalions, who, at half-past eleven o'clock, were comfortably eating their dinners. In less than a minute hundreds of these fellows were killed, and the rest were flying eastwards in wild disorder. Next moment the Maxims [they were actually Hotchkisses] were switched on to the Russian firing line who, with their backs to the river and their attention concentrated on Penchiho, were fighting in the trenches about half-way up the slope of the mountain. These, before they could realise what had happened, found themselves being pelted with bullets from the rear. No troops could stand such treatment for long, and in less than no time the two brigades of Russians which had formed the extreme left of Stakelberg's attack, were in full retreat. Altogether the six Maxims [Hotchkisses]

had accounted for, according to the first despatch, 1000; according to the second, 1300 Russians. (Hamilton 1907: II, 239–40)

Gradually, both sides realized that the machine gun was as useful in mobile roles as it was in static positions and tactics evolved accordingly. By the Battle of Mukden, the greatest land battle that had been fought prior to World War I, there had been a huge increase in the issue of machine guns. It was not unknown for 24 Japanese machine guns to be grouped together.

THE HOTCHKISS IN MEXICO

The Mexican government had been among the first to purchase Hotchkiss machine guns in quantity, buying small quantities of the original 1897 pattern to replace an assortment of manually operated Gatling Guns. The French guns were accompanied by the original type of lightweight tripod, lacking any provision for traverse though elevation could be adjusted by a small handwheel. At least one commentator (Hughes 1968: 98) has recorded that marks on the machine-gun bodies suggested that the weapons had been made not by Hotchkiss & Cie in Saint-Denis but instead by Compagnie des Hauts-Fourneaux, Forges et Aciéries de la Marine et des Chemins de Fer in Saint-Chamond.

An explanation that the Hotchkisses were surviving French field-trials guns which had been assembled in Saint-Chamond using Hotchkiss parts (cf. the links between Springfield Armory and Colt when the Benét-Mercié Machine Rifle was being produced) overlooks the fact that the Saint-Chamond factory was privately owned. It seems

The M1900 Hotchkiss served throughout the Mexican Revolution (1910–20). One example is shown here with government soldiers who are also firing M1895 Mauser rifles. Note the early type of tripod, with provision for elevation but not traverse. (Paul Scarlata)

TOOLS AND ACCESSORIES

The French M1914, like most other Hotchkiss machine guns, was issued with a bewildering selection of tools, spare parts and cleaning equipment. According to the handbook, the *Sac à chiffon* ('rag bag') contained about 1kg of wiping cloths, a *seau en toile* (canvas bucket), an *épaulière* (padded shoulder protector) for use when carrying the gun, and a pair of heat-resistant gloves to be used when changing a hot barrel.

The *Petite sacoche* (small bag or satchel) contained the most immediate needs: a replacement extractor and its spring, a firing pin, an oil bottle, a *crochet-ejecteur* (a hook-like manual ejecting tool), a spanner-like gas regulator key, a defective-case extractor known as the Tire-douille M1907, and a small brush.

The *Grande sacoche* (large bag), usually kept in the rear, contained an M1908 barrel-removal key and the key needed to dismantle the tripod, a *crochet-ejecteur*, a multi-part cleaning rod and six brushes, two brushes and a scraper to clean the gas cylinder, an oil bottle, a tin of lubricating grease, a small rasp, a hammer and a screwdriver. There were also usually two *Niveaux*

A British .303in Hotchkiss Mk I* No. 2, no. E.28477, with an ammunition box and a non-matching transit chest for E.39127. (Rock Island Auctions, www.rockislandauction.com)

much more likely that an association between the Mexican arms-designer Manuel Mondragón who, with the co-operation of Colonel Émile Rimailho, the technical director at the Saint-Chamond factory, produced the 75mm Saint-Chamond-Mondragón field gun in the 1890s, led to a concurrent purchase of the Hotchkiss machine guns – but that the order was placed through the Saint-Chamond factory and the Hotchkisses were marked accordingly.

The Hotchkiss machine guns served throughout the Mexican Revolution, a complicated series of regime changes that began in 1910 and lasted for more than a decade. Ordnance affairs became complicated, as weapons were acquired, officially and unofficially, from a variety of sources. Consequently, in addition to Maxim, Vickers and Madsen machine guns used by the Mexican Army (and also by some of the rebels), at least two other types of Hotchkiss saw combat: 6.5mm Kulspruta m/1900 machine guns acquired from Sweden, and .303in British Portative-type guns. The Swedish 1897-pattern machine guns were made by Hotchkiss & Cie and are assumed to have been sold to Mexico once the Hotchkiss had been replaced in Swedish service by the Schwarzlose medium machine gun; the British guns, which could not have been made prior to official adoption in 1916, were probably war-surplus purchased after 1918. However, confirmation is lacking.

de repérage (sight adjusting spirit-level clinometers) in leather pouches. Guns with the 2,000m tangent-leaf sight were issued with the M1907 Saint-Étienne corrector, but those with the 2,400m dial sight used the M1918 corrector with screw-pillar elevator. Also included in the Grande sacoche were spare parts ranging from extractors, firing pins and springs to ejectors and muzzle stoppers. Sight protectors and *Bouchons de tir à blanc* (BTB: blank-firing attachments) were carried, with a solitary *Verificateur de feuillure* (aim-corrector) per machine-gun section.

A special bracket was issued to support the *bande articulée*, attaching to the left side of the gun (land service only). There were two types of flash-hider, the older being a closed-top design with distinctive down-curving flanks while the newer pattern, more conventional in design, was simply a cone drilled with small holes.

The handbook for the British version of the Portative listed the accessories carried with the gun as a dismounting wrench, an ejector key, a manual extractor, a multi-part cleaning rod, a gas-cylinder cleaning tool, two wire brushes (the picture accompanying the original text actually shows a brush and a jag), an oil can, a cleaning brush, and a front-sight cover. A wide range of spare parts and additional equipment, such as a spare barrel and a cartridge-strip filling machine, accompanied the wheeled limber.

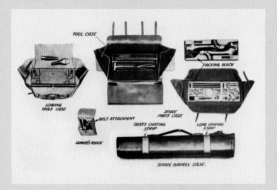

Some of the Benét-Mercié Machine Rifle's accessories, from the US Army handbook. (Author's Collection)

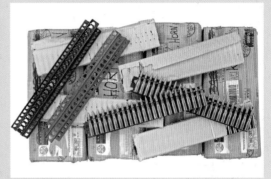

Hotchkiss ammunition feed strips, empty and loaded, atop modern 8×51mmR ammunition. (Morphy Auctions, www.morphyauctions.com)

The Mexicans also played a part in discrediting the US Army's Benét-Mercié Machine Rifle. On 9 March 1916, the Mexican revolutionary Francisco 'Pancho' Villa had ordered a raid across the US border to attack the town of Columbus, New Mexico. The raiders captured part of the settlement and the US Army base there before the defenders were able to launch a counter-attack. During the ensuing battle, all four M1909s of the 13th Cavalry's machine-gun troop saw action, each firing more than 5,000 rounds in the darkness at targets lit up by buildings on fire (Scarlata 2019). The invaders subsequently retreated across the border, leaving part of Columbus in ruins, with the cavalrymen snapping at their heels. Subsequently, basing their views on claims that the M1909 Benét-Mercié Machine Rifles were difficult to load in the dark, journalists assumed that firing lapses were due to jams instead of an absence of targets. Consequently, the M1909 Benét-Mercié Machine Rifle was disparagingly labelled the 'Daylight Gun', and the US Army was advised to partake in battle only in daytime.

This unwanted reputation persisted even though investigations by Julian Hatcher, then a captain in the US Ordnance department but later renowned as the author of *Hatcher's Notebook*, revealed the underlying problem to be a lack of even rudimentary training. There had been loading problems, attributable to lack of experience of night-firing, but the guns were believed to have fired 20,000 rounds with only an occasional jam.

THE HOTCHKISS ENTERS WORLD WAR I

When hostilities commenced in July 1914, few of the participants had extensive inventories of machine guns and only a few manufacturers were actively involved in supplying such weapons. Stories were circulated that the Germans had 50,000 machine guns, when there were probably only about 2,500 Maxims in front-line service at the outbreak of hostilities. The Belgian Army had a mere 102 machine guns; and, largely due to the dominance of national pride over practicality, the French and the Italian armies had been issued with machine guns of highly questionable utility.

Ignoring the many obsolete machine guns that were still being kept in fortifications or the reserve, the British had the Vickers, the Germans had the Maxim, the Austro-Hungarians had the Schwarzlose, the French had the M1905 Puteaux and the M1907 Saint-Étienne, and the Belgians had the Maxim and the Hotchkiss.

The popular view that the fighting would be over by Christmas 1914 soon faded away as 1915 brought only ever-increasing ferocity. The combatants soon learned that the machine gun was capable of prodigious slaughter, yet British, French and German generals were still sending infantrymen into withering machine-gun fire in 1916, their advance sometimes being halted before they even reached the front line. Once the value of the machine gun in static emplacements was well established, even in the trenches of the Western Front, progressive thought turned to weapons which could accompany raiding parties. The prime criterion was light weight, even though the metallurgical knowledge and production techniques of the day struggled to provide anything that combined portability with durability.

The Hotchkiss Portative had been serving in some numbers in August 1914, though confined principally to the French and the Belgian armies (and, of course, the initially non-combatant United States), though its weaknesses – principally the strip feed – ensured that it was not to be the ultimate answer. Though sometimes seen by the French authorities as a direct replacement of the Portative, the Chauchat light machine gun, a remarkable precursor of the cheap and easily made guns developed after 1918, was let down by poor metallurgy and an inattention to detail.

THE HOTCHKISS M1914

The French, as short of machine guns as virtually everyone else, ordered the perfected Hotchkiss to supplement the M1907 Saint-Étienne. The finalized version of the basic Odkolek design, the Mitrailleuse Hotchkiss Modèle 1914, made by the Hotchkiss manufactories in Saint-Denis, Levallois-Perret and later Lyon, chambered the French 8×51mmR cartridge. Propellant gas tapped from the mid-point of the barrel drove the bolt backward as the gun fired, allowing cam surfaces on the carrier to pivot the locking struts out of engagement in the body walls.

Though similar externally to the M1900, the M1914 version embodied several improvements. The feed block and the pistol grip were still bronze castings, but the feed was refined, the M1900-type safety catch was

THE HOTCHKISS AND THE FORTS

In 1914, the French, like many continental powers, had long-established defensive lines equipped with turrets or embrasures mounting large-calibre cannon, to strike at long-range targets, and machine guns to provide local defence.

A special disappearing turret had been developed in the 1890s, apparently by one Lieutenant-Colonel Bussière with the assistance of manufacturers Compagnie des Fives-Lille and Châtillon & Commentery. Consisting of an iron cylinder topped by an armoured cupola, set in reinforced concrete and operated by a counterweight mechanism, Tourelle GF-3 held a single seven-barrel 8×51mmR Gatling Gun. When the Hotchkiss was approved, however, Tourelle GF-4 appeared with two M1900 machine guns mounted *en echelon* on a sliding carriage so that they could be used simultaneously.

A safety system was introduced in 1902 to prevent the turret being lowered while the gun muzzles were still protruding, and nearly 90 installations had been completed by the summer of 1914. They included virtually all the best-known fortifications, including the ring around Verdun. Fort Douaumont had two turrets, destroyed during the German bombardment in 1916 but subsequently replaced.

Usually capable of an elevation of 8° and a depression of 9°, many *tourelles à l'éclipse* (disappearing turrets) survived until sent for scrap in 1943 during the German occupation. However, most of them had already been re-armed with 7.5mm MAC 1931 machine guns.

A view of two disappearing turrets, showing how the two Hotchkiss machine guns were placed one above the other, *en echelon*. (Author's Collection)

abandoned, and the regulator was restricted to four positions (marked '1' to '4', '3' being considered the normal setting) instead of 20. The barrel of the M1914 was much easier to change than the M1900 type, being held by an interrupted screw-thread, and a special key was provided to make the change even simpler. The change was not new, as it had been protected by French Patent 375307 which had been granted several years earlier (INPI 1907). The machining of the body, in particular, was simplified; and the M1900-type piston was replaced by a reinforced design capable of withstanding 90,000 shots. Consequently, the spare pistons that had accompanied the M1900 Hotchkiss were no longer needed.

A well-trained M1914 Hotchkiss machine-gun crew could sustain fire for surprisingly long periods, as long as care was taken not to damage the ammunition feed strips. Fortunately, the mechanical strip-filling machine (INPI 1912) also re-set any of the cartridge-holding fingers that had been bent over. A manual strip-reforming tool was another essential part of the accessory kit that accompanied each gun. The only other problem arose from the ammunition itself; attempts were made initially to restrict the M1914 to the Balle 1886 M, as the primers of the Balle 1886 D had proved to be unsatisfactory in the M1907 Saint-Étienne. The introduction of Balle 1886 D a.m. (*amorçage modifié*: 'altered priming') ammunition in 1915, with different primers, corrected the problem.

The Hotchkiss served in practically every theatre of World War I, from the First Battle of the Marne (6–12 September 1914) to the Armistice on 11 November 1918. The fighting around Verdun in 1916 was particularly intense, and very costly in both lives and equipment. When the Germans

M1914 Hotchkiss no. 36936 of 1918, on the M1915 Type Omnibus tripod. (Morphy Auctions, www.morphyauctions.com)

MOUNTS

The M1897-type tripod was a very simple design, lacking traverse and with only the most basic elevation control. Though light, it proved to be ineffectual even though the Swedish ksp/00 had a wheel on the rear leg to improve mobility. Hotchkiss & Cie seems to have offered the M1900 tripod in two versions: simple (with limited adjustability) and a more robust version with a traverse system built into the tripod body. The latter proved good enough to remain in production until World War I began and the M1914 *Type Guerre* was substituted. The two-position M1914 tripod weighed about 23.5kg, but was replaced in turn by what is customarily known as the M1916, though protection for this 25kg three-position design was not sought by Hotchkiss & Cie until 23 August 1917. French Patent 503684, *Système d'affût-trépied pour mitrailleuses*, was not accepted until 23 March 1920. The M1916's traverse system, with the limit-stops placed at their extremities, gave a radial movement of 37° either side of the centreline.

Hotchkiss machine guns could also be mounted on the M1907 Saint-Étienne and M1915 Type Omnibus tripods if appropriate adaptors were available, the M1915 being distinguished by its somewhat skeletal appearance compared with the Hotchkiss designs, its greater height above the ground, and a large handwheel on the left side of the body. The front legs could be folded to reduce the tripod's overall silhouette.

M1914 Hotchkiss no. 27265, dating from 1917, on the US Army's Standard Products Company tripod, with 360° traverse. (Rock Island Auctions, www.rockislandauction.com)

Standard Parts Company tripod mount maker's plate. (Morphy Auctions, www.morphyauctions.com)

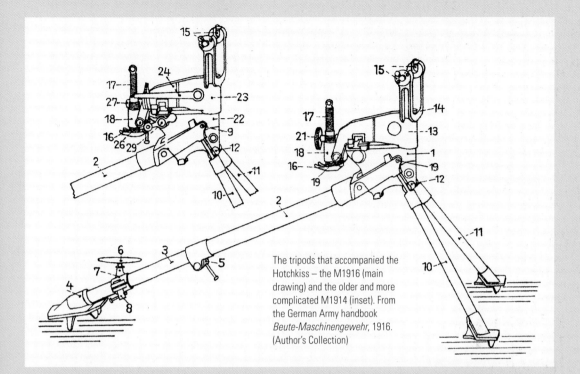

The tripods that accompanied the Hotchkiss – the M1916 (main drawing) and the older and more complicated M1914 (inset). From the German Army handbook *Beute-Maschinengewehr*, 1916. (Author's Collection)

Some of the Hotchkiss machine guns shipped back to the United States for training purposes were fitted to the distinctive M1916 tripods made by the Standard Parts Company of Cleveland, Ohio. These were generally comparable with the Hotchkiss patterns but had a unique 360° traverse system which cannot be mistaken for any other mount.

Among the special-purpose mounts were the *Affût de Rempart Modèle 1907*, based on a French patent sought by Hotchkiss & Cie on 2 February 1904 and accepted in June that year as French Patent 340111. A rack-and-pinion system enabled the machine gun to move up the sturdy tubular support to fire, or down to safety when required. A slender conical pillar, often fitted with an armoured shield, was used in fortifications and aboard ships.

The *Affût contre-avion 'Jean'* comprised a mounting bracket on top of a tube, 1.9m tall, which was placed vertically on its foot and stabilized by stakes driven into the ground through the tips of three rod-like legs. Post-war designs included the *Affûts Contre-Avion* ('CA') M1925 and CA D M1926, the latter with two guns superimposed and slightly staggered so that their belt-feed ammunition drums did not interfere with each other. A special Cazaut-Labat anti-aircraft sight was issued with the M1914 Hotchkiss, consisting of a fixed bracket for the *mire* (rear sight) and a removable barrel-mounted collar for the *guidon* (front sight). Eventually, a special extended bracket, the *Rallonge de tir contre-avion* M1928, was produced to fit the M1915 Type Omnibus mount. It was usually accompanied by a detachable shoulder-stock, and by stadia used to estimate range.

The three-positional Hotchkiss M1916 tripod. (Author's Collection)

Two views of the standard M1914 Hotchkiss, no. 23069 of 1917. The M1914 Hotchkiss was 1,310mm long, had a 785mm barrel, and weighed about 23.5kg. The M1915 Type Omnibus tripod added 24.0kg. The ammunition feed strips each held 24 rounds, though after 1916 the articulated metal belt developed primarily for use in the air, made of three-cartridge sections (maximum length 249 rounds), could be used. Cyclic rate averaged 600rd/min. (Morphy Auctions, www.morphyauctions.com)

were finally forced back, casualties in the February–December 1916 period alone, according to one recent estimate (Heer & Naumann 2000: 26), amounted to 377,231 French and 337,000 German with 40–45 per cent killed or missing in action.

Combat experience, particularly in the trenches of the Western Front, revealed the shortcomings of the M1907 Saint-Étienne machine gun. An improved M1907/16 appeared with a new gas-pressure regulator, alterations to the firing pin, and a new 2,400m rear sight, with a dial-type base, replacing the original tangent-leaf sight. However, the modifications were too few, and too late. The M1907/16 offered so little improvement on its predecessor – even though thousands were made, peaking at about 1,900 per month by January 1917 – that the decision was made to concentrate on the M1914 Hotchkiss.

Deliveries of the M1914 began in December 1914, production rising to about 1,400 per month by February 1917 and peaking at about 2,000 per month in July 1918. Only 100 M1914 guns had been made in 1914, but production rapidly accelerated – 2,300 in 1915, 9,350 in 1916, 17,200 in 1917 and 16,900 in 1918. Numbered from 1 to 46850, M1914 guns usually bore their maker's name on the left rear side of the body. The designation, serial number and date were clustered on the right side of the body above the trigger, originally as 'M1914' above '5678' over '1916', though guns made in 1918 often omitted the model designator. Serial numbers were repeated on the barrel, below 'H' for Hotchkiss, and on many of the parts. The letter 'M' on the feedway indicated guns that had been adapted to accept *bandes articulées*, 'N' on the barrel showed alterations had been made for the Balle 1932 N, and a large 'X' identified those guns which had been reduced to instruction or drill status.

Work on the M1907 Saint-Étienne declined as rapidly as M1914 Hotchkiss output rose; by May 1918, monthly deliveries of the M1907/16 were averaging only about 50 guns. Many Saint-Étiennes were scrapped after the Armistice, but some, altered to accept conventional fabric ammunition belts, were held in store for many years; others were despatched to the French colonies.

The basic M1914 Hotchkiss was enlarged in 1915 to create the Balloon Gun. Chambered for 11mm Gras ammunition loaded with

CREWING AND SUPPLY

The complexity of early machine guns necessitated the services of more than one man. By 1905, for example, each infantry division of the Japanese First Army had four batteries of six tripod-mounted Hotchkisses, each battery being manned by an officer and 74 men.

In French Army service, and in most other armies, a Hotchkiss would have had a crew of four: the *chef de pièce* (gun commander), responsible for selecting targets, field of fire and carrying out instructions of the officer or senior NCO of the unit; the *tireur* (firer), who engaged the target when instructed to do so and was responsible for maintaining the gun when time permitted; the *chargeur* (loader), whose task was to feed strips of ammunition into the gun; and the *aide-chargeur* (loader's assistant), who brought up ammunition from the rear and defended the gun and its crew when appropriate. The men usually carried Berthier mousquetons (carbines) for self-defence.

A British 'Light Hotchkiss' machine-gun section comprised five men – No. 1, a junior NCO, was the firer; No. 2 was his assistant, who fed the gun and watched for signals from officers or senior NCOs; No. 3 was the ammunition carrier, usually positioned a few paces to the rear; No. 4 was the limber-man, ready to bring more ammunition up to No. 3 when requested to do so; and No. 5 acted as a scout, range finder or sniper to protect the gun and its crew.

The French M1908/13 Portative was also regarded largely as a cavalry weapon. The gun could be carried on the pommel of a special saddle, retained by studs and quick-release straps. Another horse carried the *selle de munition* (ammunition saddle), with the spare gun-barrel and two *sacoches* (bags or satchels) each containing two *paquets* (packages) of 12 feed strips apiece: a total of 720 rounds. The *bât de pièce* (gun pack) comprised two Fusils Mitrailleur, replacement barrels, cleaning equipment and accessories, and two to four *coffres de munition* (ammunition chests), each containing 300 rounds. A maximum load of 120kg was permitted. The *bât de munitions* (ammunition pack), another horse-load, contained 1,800 rounds in six 300-round boxes.

Members of a French Army Hotchkiss machine-gun crew providing fire-support for British infantrymen take a break, somewhere on the Western Front. The steel helmets suggest that the photograph was taken in 1916 or later; and their lassitude suggests that the men are not under fire. (Author's Collection)

incendiary bullets, this weapon was intended to bring down the German observation and artillery-spotting balloons that customarily rose out of rifle range behind the front line. The success of these large-calibre guns is said to have inspired development of not only the .50 Browning heavy machine gun but also the 13.2mm Hotchkiss machine guns of the 1930s.

Acquisitions of M1914 machine guns by the US Army totalled 9,592, commencing almost as soon as the American Expeditionary Forces arrived in Europe in the summer of 1917; 1,300 of these guns were then shipped back to the United States for training purposes, often to be fitted with M1916 tripods made by the Standard Parts Company of Cleveland, Ohio. Hotchkiss guns also served with British Army units, particularly in districts bordering or shared with the French Army.

In addition, Hotchkiss machine guns were among those given to the Italian Army to make good losses after the disastrous Battle of Caporetto (24 October–19 November 1917) was perceived to have given the Austro-Hungarians and their German reinforcements an upper hand. Large numbers of M1914 Hotchkiss machine guns captured on the Western Front, in particular, were not only pressed into service with the Central Powers but also included in instruction manuals such as the German *Beute-Maschinengewehre* ('Captured machine guns', 1916) alongside the British Colt, Lewis and Vickers guns, the French Saint-Étienne and the Russian Maxims.

THE HOTCHKISS IN THE AIR

The first practical experiments with machine guns in the air were undertaken in the United States at the beginning of June 1912, with a prototype Lewis Gun. However, the idea of arming aeroplanes was much older. At the Paris Salon de l'Aéronautique of 1910, for example, a 37mm Hotchkiss single-barrel navy cannon (not the multi-barrel revolving type) had been displayed mounted in the nose of a Voisin two-seater biplane.

THE HOTCHKISS AND ARMOURED VEHICLES

An M1909 Benét-Mercié Machine Rifle mounted on a Harley-Davidson motorcycle. (Author's Collection)

The idea of mounting machine guns on vehicles was nothing new. The Gatling, Gardner, Nordenfelt and others had often been placed on wheeled carriages, to be towed by horses as a form of light artillery. A few visionaries had mounted machine guns on carts propelled by bicycles, and, when the internal-combustion engine led to the automobile, it was but a short step to truly self-propelled guns.

Among the exhibits at the Salon de l'Automobile which opened in Brussels on 8 March 1902, therefore, was the Charron & Girardot-Voigt armoured car. This consisted of an open-bodied two-seater with an M1900-type Hotchkiss, protected by an armoured shield, mounted on a pillar inside a tub-like armour-plate superstructure. A tripod was carried on the outside of the vehicle behind the driver's seat to allow the M1900 Hotchkiss to be dismounted when required.

Exposing the driver and his passenger to hostile fire was entirely inappropriate, but, before long, enclosed-body designs had been proffered alongside light machine guns such as the M1908 Portative carried in motorcycle sidecars.

The first use of vehicle-mounted Hotchkiss machine guns in combat occurred when German forces invaded Belgium in August 1914. Inspiration is usually credited to Charles Henkart, an officer-reservist who, when he re-joined the Grenadiers on mobilization, brought with him two Minerva touring cars which had been fitted with Cockerill armour-plate bodies and a single 7.65×53mm M1900-type Hotchkiss firing rearward. The pillar-mounted gun had an armoured shield to protect the gunner and was accompanied by 4,500 rounds of ammunition in 30-round strips.

The Minerva armoured cars had a four-man crew, weighed about 4 tonnes, and could reach 40km/h on made-up roads. Combat experience soon revealed their potential, particularly when used against German infantry advancing without artillery support, and contributed to some of the few Belgian successes against the German forces. A Corps of Armoured Cars was formed in August 1914 and, when Belgium finally capitulated in October 1914, at least 30 armoured cars had been built in Antwerp ('Anvers' in French) by Société Anonyme Minerva Motors and Société Anversoire pour Fabrication de Voitures Automobiles (SAVA).

When the first tanks appeared in 1916/17, ponderous and slow-moving, it was only natural that they mounted machine guns to provide short-range defence. The French Schneider CA series and Saint-Chamond tanks, in addition to their 75mm main guns, carried two and four M1914 Hotchkiss machine guns respectively. The Renault light tank initially had a single M1914-type Hotchkiss even though the design of the turret was altered several times. One, two or even more Hotchkisses (usually the M1914, but occasionally Portatives) could be found on Peugeot, Renault and other French armoured cars.

Published in 1915 by Léopold Verger & Cie of Paris, this postcard honoured Belgian Army officer Lieutenant Charles Henkart, who, with the crew of his Minerva armoured car, was killed in September 1914 after an epic two-hour battle with German cavalrymen. (Author's Collection)

The Belgian SAVA armoured car carried a single Hotchkiss gun in a turret on the right rear of the body-shell. Like the Minerva, the SAVA had a cladding of Cockerill armour plate. (Author's Collection)

This was little more than a publicity gimmick, however, as the limited power of the Voisin's engine was scarcely enough to get the aeroplane, its crew and fuel load into the air even without the considerable additional weight of the gun. It was, however, a pointer to the future.

Many inventors were drawn to the idea of aerial combat, their vision often at odds with military conventions which saw aeroplanes as fit only for scouting and reconnaissance. At the 1913 Paris Salon de l'Aéronautique, exhibits included a Nieuport armoured monoplane equipped with a Hotchkiss Portative machine gun mounted on a pillar. In August 1913 the French authorities issued specifications for *avions de combat*, and a large twin-engine biplane developed by Commandant Émile Dorand had been shown to General Joseph Joffre on 6 June 1913. This also had a Hotchkiss Portative mounted in the nose of the central nacelle. At much the same time, Hotchkiss & Cie published a promotional booklet which included photographs of the Portative mounted on Farman aircraft.

More successful was the TT monoplane developed by Armand Deperdussin but said to have been designed by his engineer Pierre Loizeau. French Patent 475080, accepted on 16 January 1914, shows how a gunner in the front cockpit could stand to fire a Hotchkiss Portative fitted to a tubular frame over the propeller arc. Two prototypes were demonstrated at Villacoublay in July 1914, on the eve of war, and there is evidence to show that similarly armed monoplanes saw service when World War I began. The arrangement looks to be clumsy, but, according to Loizeau, had no adverse effect on the aeroplane's stability. However, the gunner undoubtedly blocked much of the pilot's view when firing the gun.

Aerial combat showed the value of machine guns, but it also revealed that fixed forward-facing guns, unless angled upward (and thus difficult to aim), were prone to shoot through the aircraft's propeller blades. This problem was partly solved by fitting steel deflector plates to the customarily wooden propellers – credited to Roland Garros, who had modified his Morane Type N monoplane appropriately – but the real

THE FIRST AUTHENTICATED AERIAL KILL

The first shots in aerial combat occurred in November 1913, during the Mexican Revolution (1910–20), when two American mercenaries fired pistols at each other, without, it is said, any intent to kill. On 5 October 1914, however, Voisin III no. V 89 – one of the six armed biplanes that had been donated by the manufacturer – piloted by Sergeant Joseph Frantz encountered German Aviatik B.I reconnaissance biplane B.114/14, flown by Leutnant Wilhelm Schlichting, which was scouting above the Marne. Frantz's observer, Mécanicien Louis Quenault then fired 47 rounds in several short bursts from his Hotchkiss Portative at his opponents. Fatally wounded, Schlichting lost control. The Aviatik plunged to the ground, killing observer Fritz von Zangen as it crashed near Jonchery and burst into flames. War in the air would never be the same again.

Though considerable use was made of Hotchkiss machine guns in the air, their salient characteristics – gas operation, strip feed – proved to be far from ideal. Though the 11mm Balloon Gun (converted from old 8mm Hotchkisses to handle Gras ammunition) was sometimes mounted in the nose of Farman HF.20 reconnaissance aircraft so that incendiary-bulleted ammunition could be used against observation balloons and Zeppelins, the Portative was preferred to even stripped-down M1900 or M1914 Army-type Hotchkisses largely because it was considerably lighter and easier to manoeuvre in the air. Some modern artworks show the Voisin III of Frantz and Quenault armed with an M1914 Hotchkiss in the nose, but the M1914 had not been introduced to service when that combat took place; there is no doubt that the weapon was a Portative.

solution was to be found in synchronizing the gun and propeller so that the former fired only when the latter's blades were clear of the muzzle.

The idea of gun–propeller synchronization is widely credited to Anthony Fokker, largely owing to self-promotion and some uncritical biographies, but the first patent was granted in Germany on 15 July 1913 to Franz Schneider of Johannisthal bei Berlin. German Patent 276396 showed a machine gun, mounted over the cylinders of an inline engine, operated by a propeller-driven linkage. Where Schneider led, others followed. In France, Deperdussin, by then in jail for embezzlement, proposed what was to become French Patent 475151 of 22 January 1914 to protect ideas for electrical or mechanical synchronization; and, on 14 April 1914, Raymond Saulnier was granted French Patent 470838 to protect a method of firing a machine gun with an oscillating rod driven from the oil pump mounted at the rear of a rotary engine. Drawings show a Hotchkiss Portative mounted above the engine cowling on a tubular frame. Tests (Woodman 1989: 23) showed that the system worked well enough to be committed to battle, but rival designs were to prove more effectual.

The one-piece metal feed strip was supplemented by a skeletal drum (INPI 1915), known colloquially as the Bobine, which supported a 75-round ammunition belt made from 25 three-round striplets hinged together. An extended ring-headed tab eased the problem of pulling the belt into the gun for the first shot.

The Portative had too many flaws to be a successful aircraft machine gun. Yet from the beginning of 1915 until superseded by the Lewis Gun in the

Jules-Charles-Toussaint Védrines, the cigarette-smoking pilot of this Morane-Saulnier monoplane with deflectors fitted to the propeller, was the first man to fly at 100mph (161km/h). He also claimed to have shot down a Rumpler Taube monoplane on 2 September 1914; though commemorative postcards had been printed, independent corroboration of what would have been the first air-combat kill is lacking, however. (Author's Collection)

middle of 1916, it armed aeroplanes including the Farman F.40, the Nieuport Type 10 and the Voisin LA3. The Farman carried a Portative on a fixed-ring mount in the nacelle nose; the Nieuport had a hole in the centre section of the wing above the crew compartment, through which the gunner could reach a Portative mounted on a pillar; and the Voisin had a pillar mount on a tripod or quadruped frame above the pilot's head.

Delivery of Lewis Guns was initially slow – records show that only 696 had been given by the British Ministry of Munitions to the French by 31 December 1916 – and so Portatives soldiered on well into 1917 before being returned to land service.

THE HOTCHKISS AT GALLIPOLI

Early in the 20th century, French advisors – including André Berthier – were employed to modernize the Turkish arms industry. Fifty M1900 Hotchkiss machine guns are said to have been ordered by Turkey in 1906, but the order was almost immediately cancelled when French influence gave way to German. However, 70 Hotchkiss machine guns were subsequently delivered, perhaps at a time when German Maxim production was focused on the needs of the Imperial German Army and the Imperial German Navy. It is assumed that the 70 guns chambered the Turkish 7.65×53mm rifle cartridge.

In addition, essentially similar guns armed the Turkish warships that had also been ordered in France: nine small gunboats of the Taşköprü class, five of which were sunk by the Italian Navy on 7 January 1912, and the three gunboats of the Îsâ Reis class. These vessels are believed to have each carried two 7.65mm M1900-type Hotchkiss machine guns in addition to their main armament.

When World War I began, each Turkish Army infantry regiment was supposed to have four machine guns, but automatic weapons were in such short supply that issue was never sufficient. Hotchkiss machine guns are known to have been used by Turkish forces opposing the Allied landings at Gallipoli in April–May 1915, but there is no evidence to show how many, if any of the 70 machine guns acquired c.1907 were employed in this particular campaign. At least three Hotchkisses opposed the ANZAC landings, but they could just as easily have been taken from the Turkish gunboats.

Bootsmannsmaat Paul Kubrille of the supply ship *Corcovado* related how the machine-gun team of Oberleutnant zur See Boltz, raised from the German battlecruiser *Goeben* and light cruiser *Breslau* temporarily transferred to the Turkish Navy (though under the command of German Konteradmiral Wilhelm Souchon), had been armed with six Maxims and two Hotchkisses. Another witness, Emile Skora of the command ship *General*, and himself subsequently attached to the Independent [German] Machine-gun Section, testified that on 3 May 1915 at Domuz Dere, eight men had been killed and several wounded – including Boltz and his second-in-command Fähnrich zur See von Rabenau – and two machine guns had been lost (Ewen 2017: 46). Skora also mentioned that the

Turkish forces employed a 'French' machine gun – presumably a Hotchkiss – near his position.

Several Australian soldiers have also testified to the presence of Hotchkiss machine guns. Corporal Joseph Wetherall of the 10th Battalion noted that the crew of a Turkish machine gun had removed their weapon from its tripod mount and thrown it over a cliff as the Australians advanced; and Major Drake Brockman gave an interview in which he thought a Hotchkiss had been captured on MacLagan's Ridge by the 10th Battalion.

THE BRITISH EMPIRE AND THE HOTCHKISS

The British Army made limited use of M1914-type Hotchkiss machine guns during World War I, most notably in sectors where British units were operating under French command, but not in sufficient numbers to displace the Vickers Gun from its pre-eminent role.

However, the Hotchkiss Portative was officially adopted by the British Army in June 1916, largely because it was thought to be better suited to cavalry use – and easier to manufacture – than the Vickers Gun. It has often been claimed that the Hotchkiss Portatives were made at the Royal Small Arms Factory, Enfield, but this is untrue: they were made in Coventry. Mystery still surrounds the activities of Hotchkiss & Cie in Britain, but, most probably, worries about the German threat to Paris (which led to a partial relocation of the Saint-Denis and Levallois-Perret manufactories to Lyon) also led to what was effectively the transfer of the Portative, no longer regarded as a regulation French weapon and thus something of an irrelevance.

Hotchkiss & Cie initially took over the premises of the defunct Arno Motor Co. Ltd in Gosford Street, Coventry, and built a large factory nearby, completed in 1917, which was known as the 'Artillery Works'. Coventry newspapers affirm that, far from the French production line

British Hotchkiss Mk I* No. 2, E.27497. (Morphy Auctions, www.morphyauctions.com)

Men of 2nd Queen Victoria's Own Rajput Light Infantry prepare to fire their Mark I Hotchkiss in this postcard dating from World War I. (Author's Collection)

being relocated, the machinery was new; the British Portative was made to imperial standards instead of metric, and so the gauges and tools would have been different.

Official records reveal that the first British contract had been placed on 20 November 1915 for 3,000 guns, delivery of which was to commence in March 1916. A subsequent order for 1,950 (never fulfilled, possibly owing to the upgrade to Mk I*) was followed on 1 February 1917 by a third order for 4,500 guns, and presumably by more orders in 1918. It has been suggested that output reached 40,000 or even 50,000 by the end of the war; evidence is lacking, though the E-prefixed serial numbers (for 'English', not 'Enfield' as is widely supposed) run at least to E.39267. However, no gun with a four-digit number has been found, and the official drawing, SAID 2064, can be interpreted to suggest that numbers began either at 10000 or 10001. Each gun was accompanied by two barrels, identified by 'A' and 'B' instead of the 'E' prefix.

The Gun, Machine, Hotchkiss, 0.303-inch Mark I was 46.25in (1,175mm) long, had a 23.5in (597mm) barrel, and weighed about 27lb (12.2kg) with its bipod. Thirty rounds were generally carried in each ammunition feed strip, though special short versions were made to be carried in bandoliers; muzzle velocity was 2,425ft/sec (739m/sec) with standard Mk VIIZ ball ammunition, and cyclic rate averaged 500rd/min. Light Hotchkiss guns fed from the right and were cocked by a bolt-handle protruding from the back of the receiver above the pistol grip. The handle doubled as a fire selector, depending on how far it was rotated upward after cocking the gun.

The Mark I had a wooden butt, with an integral pistol grip, an oil bottle and a hinged shoulder plate. It could only feed from conventional metal ammunition strips and was issued with a small 'cavalry' tripod. The Mark I*, accepted in June 1917, was made specifically for infantry (Mk I* No. 1) or tank use (Mk I* No. 2). The No. 1 and No. 2 guns were identical except for the sights and butt fittings. The No. 1 gun had a conventional tangent-leaf rear sight offset on the left side of the feed cover, and a wood butt with distinctive metal strengthening plates intended to overcome a weakness in the wrist of the original unreinforced butt, which tended to break away from the tangs. The No. 2 gun had a pistol grip adapted to take an optional tubular extension. Most tank guns were also issued with the 'Sight, Tubular, No. 2', giving an instant picture of a potential target, a pan for a 249-round cartridge belt – a series of articulated three-round strips – and a canvas bag to catch the ejected cases.

Though the Hotchkiss was light enough to be carried forward with the infantry during an attack, its strip feed was inconvenient. This, together with worries about its reliability, often confined the Hotchkiss to static roles on the Western Front. However, unlike the US M1909

Benét-Mercié Machine Rifle, which though taken to Europe in small numbers never saw combat, British guns regularly saw active service:

A British Hotchkiss Mk I* No. 2 showing the maker's mark (note the American-style spelling of 'CALIBER') and designation. The view of the bolt cap shows how the state of fire could be altered: 'S' for safety, 'R' for repetitive single-shot fire, and 'A' for automatic operation. (Morphy Auctions, www.morphyauctions.com)

> Award of the Distinguished Conduct Medal to Lance Corporal W.R. Grayson, of a Light Horse Regiment, 'For conspicuous gallantry and devotion to duty. After a severe fall with his horse into a shell hole, from which he was extricated with great difficulty, he went on after his detachment, carrying his Hotchkiss rifle, which he brought into action in time to stop a hostile field gun from being withdrawn. He was badly bruised and shaken, and showed a fine example of pluck and determination.' (*London Gazette*, 25 August 1917)

In the desert conditions of Sinai and Palestine, the Hotchkisses were used to good effect by the Australian Light Horse, the New Zealand Mounted Rifles Brigade and the Imperial Camel Corps. Records show that each mounted squadron, nominally 160 men strong, had its solitary Lewis Gun replaced by four light Hotchkisses in March–April 1917. The new weapons, together with a spare barrel and 900 rounds apiece, were carried by pack horses. A squadron also had two ammunition horses, each weighed down by eighty 30-round strips. Lieutenant James Nelson Stubbs of the 8th Australian Light Horse Regiment was awarded the Military Cross because:

> On the 25th September, 1918, he was sent with two sections and a Hotchkiss ... to secure the road leading north from Tiberias. This he did successfully under machine-gun fire. About 200 of the enemy, three motor-cars, two motor-lorries, and a number of horse transport were turned back to the town and captured. It was largely due to his gallant and able leadership that the enemy withdrew his machine guns and made it possible to enter the town. (*London Gazette*, 4 October 1919)

It is assumed that the Lewis Guns, effectual though they proved to be on the Western Front, were prone to jamming in desert sand. The problems were probably due more to the feed than the operating system, however, as many changes were made to the magazines to correct weak construction

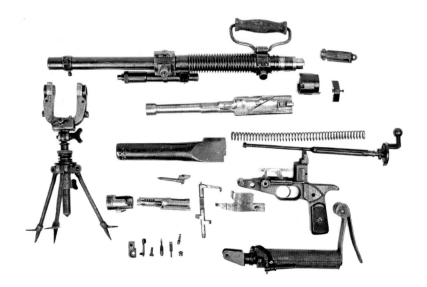

The principal components of the Hotchkiss Mk I* No. 2 and the diminutive 'cavalry tripod'. (International Military Antiques, www.ima-usa.com)

and tendencies to jam. The original land-service pattern, retrospectively designated No. 1 and then No. 1 Mk I, had a plain sheet-metal body; No. 2 (No. 2 Mk I) was strengthened with radial ribs; and No. 5 Mk I and No. 6 Mk I, converted from No. 1 and No. 2 respectively, gained holes in the cover to give access to a spring-loaded release catch.

The light Hotchkisses armed many British tanks, the design of the barrel suiting them to the role better than the Lewis Gun; the latter's large-diameter barrel casing was incompatible with the ball mounts that were

Pictured in Crete in 1941, these Australian soldiers enjoy a game of cards during a break from their anti-aircraft duties. The small 'cavalry tripod' of the Hotchkiss has been fitted to an extension and a larger tripod necessary to raise the gun above shoulder height. (Paul Scarlata)

usually preferred, as it allowed enemy bullets to penetrate. The Mark II 'Male' tank, for example, had two Hotchkiss 6-pdr QF guns and three Hotchkiss machine guns; the Mark II 'Female' tank had three Vickers and one Hotchkiss machine gun. The Medium Mark A Whippet tank had four Hotchkiss machine guns, and the comparatively few twin-turreted Austin armoured cars in British service, of a type originally made for Russia with Maxims, were equipped with Hotchkisses. Most of those Austins that survived into the 1920s, their bodies fitted to Peerless commercial-vehicle chassis, were still in service when World War II began.

Many of the light Hotchkisses were sent immediately after World War I to India, thus ensuring that their useful life was extended for many years even though they were nominally superseded first by the Vickers Berthier and then by the Bren Gun. Others, stored for many years, reappeared at the beginning of World War II.

THE US HOTCHKISS IN WORLD WAR I

When the United States entered World War I in April 1917, and the American Expeditionary Forces (AEF) received orders to depart for Europe, there was a notable shortage of arms and equipment. When an inventory was taken on 30 June 1915, the US Army possessed only 1,236 machine guns: 287 M1904 Maxims, 670 of the largely discredited M1909 Benét-Mercié Machine Rifles (the so-called 'Daylight Guns'), 148 Colt–Brownings (M1895, M1902 and M1904), and 131 assorted manually operated Gatling Guns.

Despite its perceived shortcomings, the M1909 Benét-Mercié Machine Rifle was widely regarded as the only machine gun necessary in US service until a decision had been taken to adopt the water-cooled Vickers Gun. A few British .303in examples had been extensively tested before the war in Europe began, performing well enough to convince the US Army that the design was a considerable improvement on the M1904 Maxim that had been adopted after trials undertaken in 1901–03 against the M1895 Colt–Browning and M1900 Hotchkiss. Consequently, 125 Vickers Guns were ordered from Colt's Patent Fire Arms Manufacturing Company under the designation 'Model 1915'. Another 4,000 were requested from Colt in 1916, but none had been delivered when the AEF left for Europe. The first few M1915 Vickers Guns followed in July 1917, but only 500 examples had reached the Western Front by the end of October that year as the Vickers Gun had been superseded by the M1917 Browning.

When the AEF reached Europe, therefore, the first 12 divisions were given a variety of British and French weapons – including British .303in Vickers and French 8mm Hotchkiss machine guns – and, particularly when fighting alongside British and French units, Lee-Enfield, Lebel and Berthier rifles. It seems that a few M1909 Benét-Mercié Machine Rifles had also been taken to France, but the M1914 Hotchkiss was to be the first US Army machine gun to see combat there.

The adoption of the M1917 Browning came too late for universal issue to the AEF, and so the majority of 3,125 Colt-made and 6,112

Pictured in France in 1918, this four-man machine-gun team mans a US Army M1914 Hotchkiss on an M1915 Type Omnibus mount. Note the British-style steel helmets and the M1911 Colt–Browning pistol on the gun commander's hip. (US Army)

Vickers-made M1915 machine guns on hand at the end of September 1918, and the 9,592 M1914 Hotchkiss machine guns loaned by France (1,300 of which had been sent to the United States for training purposes), were still in US Army service when the war ended. Loan guns remaining in France were returned to the United States in 1919, but those already in the United States were sold off as surplus in 1918–20.

A few Hotchkiss guns had been converted experimentally to fire the US .30-06 cartridge, apparently at the beginning of 1918. Though surprisingly successful, the time that would be lost converting weapons was judged to be unnecessarily disruptive when production of 8mm French Hotchkisses was being prioritized and work on the M1917 Browning was well under way in the United States. Colt and the New England Westinghouse Corporation had manufactured more than 66,000 water-cooled M1917 Brownings when the war ended, but only 30,582 had been despatched to Europe and fewer still had reached the front line.

AFTER THE WAR: REPARATIONS

When World War I ended, many thousands of Hotchkiss machine guns were deemed surplus to requirements. Some were scrapped, but many were sold off – the French supplied M1914 Hotchkisses and Portatives to Greece in substantial quantities – or sometimes given to the emergent nations that faced the Bolshevik horde. Some went to Romania and possibly also to Latvia, Lithuania and Estonia, the Baltic states having been carved out of the Russian empire in 1917.

Poland received large numbers of Hotchkiss guns, 2,620 examples of the cięski karabin maszynowa wz. 14 remaining on the Polish Army inventory in the mid-1930s. An attempt was made to adapt the wz. 14 to fire the 7.9×57mm cartridge – more powerful than the 8×51mmR – but the Radom-made ckm wz. 25 proved to be prone to overheating, with a consequent detrimental effect on accuracy. Replaced by the ckm wz. 30, a copy of the M1917 Colt–Browning, the Polish Hotchkisses were fitted in armoured vehicles, sold to Spain during the Spanish Civil War, or placed in store to confront the Germans once again in 1940.

The Czechoslovakian Army had 985 M1914-type Hotchkisses in 1920, most of which had been acquired from France, though some had been brought back from Russia with the Czechoslovak Legion. Most of these weapons were still available in 1939, when Germany annexed first the Sudetenland and then Czechoslovakia in its entirety.

AFTER THE WAR: JAPAN

Apart from the attack on the German protectorate of Tsingtau (Kiautshou) and the seizure of Germany's Pacific and Far Eastern territories, Japanese forces were uninvolved in World War I – but they were assiduous onlookers nevertheless. Analysis of campaigns led to the introduction in 1922 of the 6.5mm Taishō 11th Year Type light machine gun (*jūichi nenshiki keikikanjū*), whereupon the Taishō 3rd Year Type was reclassified as a heavy machine gun (*jūkikanjū*).

Generally credited to Nambu Kijirō, an artillery officer, firearms designer and founder of the Nambu Arms Company then ranking as a colonel in the Imperial Japanese Army, the 11th Year Type was probably developed by a team of technicians under Nambu's nominal control. It combined the well-proven Hotchkiss action with an unusual spring-fed open-top magazine or hopper, mounted on the left side of the body, which could be loaded with six five-round rifle chargers that acted as clips. Single rounds were then stripped into the chamber each time the bolt returned from the full-recoil position, the process being repeated until the first five-round charger was expended. The spring-loaded follower plate then ejected the empty charger through the underside of the hopper, and the first round of the second charger was fed into the breech. The process continued until all six chargers had been emptied and ejected.

Taishō 11th Year Type machine gun no. 28678, made by Hitachi Heiki in December 1939, was captured by the US Army in February 1944 on Begej Island, part of Kwajalein Atoll in the Marshall Islands. The Taishō 11th Year Type fired the special reduced-charge 6.5×50mmSR cartridge shared with the Taishō 3rd Year Type. Containing about 30.9 grains (2g) of propellant instead of the 33.2 grains (2.2g) found in the full-charge version, this cartridge developed a muzzle velocity of only 701m/sec. Ammunition was identified by a large encircled 'G' on packaging, the transliterated Japanese for 'reduced' being *genshō*. The 11th Year Type was 1,104mm long, had a 482mm barrel, and weighed a mere 10.2kg with its bipod. The butt was offset considerably to the right so that the firer's eye was positioned behind the 300–1,500m-range rear sight, which was similarly offset. Cyclic rate averaged 500rd/min. (Rock Island Auctions, www.rockislandauction.com)

The assumption had been that the hopper could be continuously replenished during lulls in firing, but reloading proved to be all but impossible on the move. This was at least partly due to the strength of the hopper-arm spring, which was appreciable, but also to the need of another hand; holding the gun, opening the hopper and inserting chargers all at once was virtually impossible.

The aim of using standard rifle chargers, simplifying ammunition supply, was undermined partly by the design of the hopper, which was complicated and unreliable, but also by the need for special reduced-charge ammunition to prevent case-head separations and associated difficulties during extraction. Regulation-issue rifle ammunition could be used in an emergency, but recoil was much more violent and extraction troubles persisted even though the gas-port valve could be adjusted (there were five positions marked '10' to '28'), and a cartridge oiler was built into the gun's body alongside the rear sight. To make matters worse, the ejector was exposed to the elements and thus prone to failure. Consequently, each weapon was accompanied by an unusually comprehensive spare-parts kit contained, with appropriate tools, in a leather satchel.

The 11th Year Type could be issued with a flimsy looking tripod, with telescoping legs which could be extended to provide an anti-aircraft mount. It also provided the basis for the Type 91 tank machine gun, accepted in 1931, with an enlarged hopper accepting ten five-round chargers. Tank guns often had a large 1.5× telescope sight on the right of the body. Shoulder stocks and bipods were carried aboard tanks and vehicles, allowing Type 91 guns to be dismounted when required, which allowed many of the surviving guns – displaced by the 7.7mm Type 97 tank machine gun – to be pressed into ground service at the end of the war in the Pacific.

The Type 89 Army aircraft machine gun was also based on the 11th Year Type. Developed in the late 1920s, when suitable machine guns were required by both the Imperial Japanese Army and Navy air forces, the Type 89 chambered a new 7.7×58mmSR Type 89 cartridge (a copy of the British 0.303in rimmed round) to avoid the problems that had arisen in the 11th Year Type with 6.5×50mm Arisaka ammunition, and had a 70-round pan magazine inspired by the Lewis Gun. A few Type 89 guns, mounted on rudimentary bipods, were even used by Japanese ground forces in the closing stages of World War II.

Taishō 11th Year Type light machine guns were made by only three contractors: Hōten Zoheishō prior to 1936, Tōkyō Gasu Denki KK (Tokyo Gas & Electric Company) until 1939, and Hitachi Heiki KK until production ended. Inspection, test firing and, if necessary, revision of individual guns by military personnel is also acknowledged in the markings on the right side of the body that have so often been misrepresented as those of the manufacturers.

Hōten-made 11th Year Types are very rarely encountered, partly because output was probably meagre and partly because surviving guns would have been taken by the Chinese in 1945. TG&E products were inspected by the manufactories in Koishikawa, from 1922 until moved to

Kokura (beginning in October 1935), and by Tōkyō Arsenal during the comparatively short time in which the Koishikawa facilities were being recommissioned in Kokura. Inspection of Hitachi guns was the responsibility of Nagoya Army Arsenal.

It has been suggested that 11th Year Type guns were made by the Nambu factory in the Kokubunji district of Tokyo, but none has ever been found; 'Nambu-made' guns usually prove to be Type 96 light machine guns made by Chuo Kogyō KK, created by merging Nambu-jū Seizosho and Taisei Kogyō in December 1936.

Output of the 11th Year Type has been estimated at 29,000 from 1922 until 1940, but confirmation is lacking. However, enough of them had been made to equip the Army and the naval landing forces long before the so-called Mukden or Manchurian incident, staged by the Japanese on 18 September 1931 by bombing the South Manchurian Railway (which was Japanese owned) and blaming the Chinese. The pretext for a long-planned invasion and annexation of Manchuria, which became the protectorate of Manchukuo, this ultimately led not only to the Second Sino-Japanese War (1937–45) but also to large-scale battles with Soviet forces in the disputed borderland of Manchukuo and the People's Republic of Mongolia.

Though some after-the-fact accounts of the battles of Lake Khasan (29 July–11 August 1938) and Khalkhin Gol (11 May–16 September

Made in December 1943 by Hitachi Heiki, Type 92 heavy machine gun no. 41454 is shown on its robust tripod, sockets on each leg allowing poles to be inserted for transport. Note the feed strips, the folding grips, and the 4× Type 96 optical sight on top of the breech. (Morphy Auctions, www.morphyauctions.com)

Kokura-made Type 92 no. 1057, completed in January 1938, with the transport poles in place in their sockets. (Morphy Auctions, www.morphyauctions.com)

1939) show the Type 96 being used in these conflicts, production of this type did not begin until late in 1937 and no deliveries were made until the summer of 1939. Consequently, the 11th Year Type was the only Japanese light machine gun serving at Khalkhin Gol. Many examples were still serving when World War II ended in August 1945.

Fighting in Manchuria in the early 1930s had shown the Japanese the advantages of the 7.9×57mm cartridge issued with the Chinese Mauser rifles, Maxim machine guns, ZB vz. 26 light machine guns, and even 7.9mm copies of the 1,192 8×51mmR M1914 Hotchkisses that had been acquired from France in the early 1930s. The power of the 7.9mm cartridge conferred not only better striking effect but also longer range, and drew attention to the weakness of 6.5×50mm Arisaka ammunition, especially the reduced-charge cartridges issued with Japanese machine guns.

The most pressing need was an upgrade of the Taishō 3rd Year Type heavy machine gun. The new 7.7mm Type 92 heavy machine gun (*kuni shiki jūkikanjū*) retained the Hotchkiss gas-operated action locked by flaps on the bolt engaging the receiver walls, but chambered a new 7.7×58mm Type 92 semi-rimmed cartridge. The lubricator was still needed to ensure efficient extraction and the original 30-round Hotchkiss feed system was perpetuated, though the ammunition feed strips were pre-loaded and could be linked together in a belt.

Production of the Type 92 was entrusted to Tōkyō Gasu Denki KK (until merged with Hitachi Seisakusho), Kokura Army Arsenal, and Hitachi Heiki KK. However, progress was slow: TG&E did not complete any Type 92 machine guns earlier than 1934, possibly beginning numbers at 10000 or 10001, and Kokura-made Type 92 no. 1057 dates from January 1937. TG&E no. 14376 dates from July 1939 and it is unlikely that numbers greatly exceeded 15000 when the manufacturing change took place in August. Hitachi Heiki continued the TG&E series, no. 24005 dating from July 1941 and 41454 from December 1943.

The Type 92 had distinctive downward-hinging traverse handles instead of the spade grips of the 3rd Year Type, and the charging slide lay on the right side of the body. It was 1,155mm long, had a 698mm barrel, and weighed 27.6kg without its tripod. Muzzle velocity was 732m/sec with standard Type 92 ball ammunition; cyclic rate averaged 450rd/min. The distinctive stuttering fire earned the gun the nickname of 'woodpecker' (or 'woodchopper') from the Australian forces which first faced it.

Type 92 heavy machine guns were often fitted with optical sights, and the updated 3rd Year Type tripod could accept a special anti-aircraft adaptor. In addition, they could fire the 7.7×58mm Arisaka Type 99 rimless cartridge, though Type 99 rifles and light machine guns would not accept the semi-rim Type 92. Japanese logistics – ammunition supply in particular – were often nightmarish.

Combat experience in the late 1930s showed that the Type 92 heavy machine gun was too heavy to manhandle easily even though sockets on the tripod legs allowed the gun to be moved with the

OPTICAL SIGHTS

Hotchkiss machine guns had comparatively conventional sights, though the original tangent-leaf pattern was eventually replaced by a sophisticated horizontal dial, operating on an eccentric to elevate the rear-sight aperture.

The US Army's Benét-Mercié Machine Rifle was one of the first machine guns to mount an optical sight, namely the Warner & Swasey 6× M1908 and then the 5.2× M1913. Forward looking in many ways, the results were uninspiring, largely because the sights could not cope with the recoil pulses. Restricted eye-relief was also a major handicap.

Most armies disregarded the value of optical sights on machine guns after World War I, the Japanese being the exception. Type 92 heavy machine guns were issued with three types of optical sight: the 6× Type 93 and the 5× Type 94 were periscope sights – 213mm and 325mm high respectively – intended for gunlaying, but the more conventional 4× Type 96 was intended to be used in combat.

Surviving Type 92 guns will often be found fitted with the Type 96 sight, but the periscopes are now rarely encountered.

The opportunity was also taken to fit 2.5× sights to the Type 96 and Type 99 light machine guns. The participation of many leading optical-instrument makers in the manufacture of these prismatic sights (necessary to ensure the line of sight bypassed the box magazine protruding upward from the body) ensured that a surprisingly large quantity reached service, though issue was far from universal. The sights were greatly appreciated by US and Australian captors, especially those who had shooting or hunting experience, as they undoubtedly contributed to the Type 96 light machine gun's reputation for accuracy. The irony, perhaps, is that the No. 32 sight now associated with British sniper rifles had been developed for the Bren Gun, not so different from the Type 96, but was never used in the fire-support role.

A Type 96 light machine gun showing the 2.5× optical sight. (International Military Antiques, www.ima-usa.com)

assistance of long poles, improving mobility in the type of terrain found in the Far East. A lighter version tested in 1940 was eventually accepted for service as the Type 1. Some of the trial weapons were fitted with a German-type sled mount, reminiscent of the old MG 08, but the finalized design had a light tripod and a readily detachable barrel inspired by trials with modifications of the Type 99. The Type 1 machine guns could only chamber the rimless Type 99 cartridge, and not the semi-rimmed Type 92. Only a few Type 92 heavy machine guns were converted to Type 1 standards, and production of the lightweight pattern does not seem ever to have got under way.

NAMBU AND HOTCHKISS: NEW DESIGNS

The Japanese forces were also in desperate need of a successor to the ineffectual 11th Year Type light machine gun, which had proved to be unreliable. Consequently, a competition was arranged in which prototypes were submitted by Nambu-jū Seizakushō, Tōkyō Gasu Denki and Nippon Tōkusō-kō. Tested in Koishikawa Arsenal shortly before the move to Kokura occurred at the end of 1935, only the Nambu design proved to be acceptable. Its rivals both suffered continual lubrication and extraction problems.

Protected by Japanese Patent 112691, sought on 6 February and accepted on 10 October 1935, the Nambu breech was locked by a vertically moving hollow locking piece in the body immediately below the leading edge of the magazine aperture (not unlike the Soviet AVS in concept), which replaced the pivoting flaps of the 11th Year Type. A quick-change barrel was introduced, and the aberrant hopper was replaced by a box magazine on top of the body. The 11th Year Type's extraction troubles persisted, however, largely because the Type 96 was still chambered for the 6.5×50mm Arisaka cartridge. The lubricator was built into individual magazines rather than the gun mechanism, removing the need for a separate reservoir in the receiver, but still encouraged dust and grit to enter the action.

The Type 96 was 1,075mm long, with a 550mm barrel, and weighed 10.2kg. The box magazine held 30 rounds; cyclic rate averaged 500rd/min. The shoulder stock and pistol grip were distinctively combined, a fixed carrying handle lay above the barrel in front of the magazine, and there was a drum-like aperture rear sight. Many, though by no means all Type 96 guns were fitted with a 2.5× optical sight. Often now defined as '2.5×13', suggesting that the diameter of the objective lens was 13mm, too small to gather light effectively, examination of surviving sights shows the marking to be '2.5× 13°' – referring to the magnification and the field of view in degrees.

A rarely encountered Type 96, no. 30191 made in Manchukuo by Hōten Zoheishō, in September 1938. (Morphy Auctions, www.morphyauctions.com)

The standard 30th Year Type sword-bayonet could be fitted around the gas cylinder but, once the barrels were fitted with flash-hiders (the earliest examples had accepted a combination muzzle protector/cleaning-rod guide), the protrusion of the sword-bayonet past the muzzle was minimal; the guns themselves, however, were sufficiently light to be used for the old French tactic of 'assault-at-the-walk'.

Series production of the Type 96 seems to have begun towards the end of 1937, but deliveries were initially slow. Though claimed to have been used in the Second Sino-Japanese War and during the confrontations with the Soviet Union in 1938–39, the Type 96 did not enter combat until the Japanese invasion of Malaya and the Philippines in December 1941. Production was initially confined to Chuo Kogyō Kabushiki Kaisha, the 1936 amalgamation of Nambu-jū Seizakusho and Taisei Kogyō, but the guns were inspected and regulated by Kokura Army Arsenal. Chuo Kogyō probably made about 10,000 of them, no. 6664 dating from October 1943.

Kokura Army Arsenal itself began production in 1938 and continued to assemble the Type 96 even after the first batches of the improved Type 99 light machine gun were made in the summer of 1942. Numbers on Kokura-made Type 96 guns reached at least 53603, dating from March 1943.

Total Type 96 output is claimed to have been 41,000 in 1936–43, but there is considerable doubt about the extent of production in the Hōten Zoheishō in Manchukuo. Claims have been made that the serial numbers of Kokura-made guns were 1–10000 and from 40000 upwards, the apparent 30,000-gun gap being allocated to Hōten. However, it has been concluded that output in Manchukuo may have been fewer than the 10,000 that is sometimes claimed. The only Hōten-made Type 96 that has been traced, no. 30191, dates from September 1937. Guns allegedly made by Nagoya Army Arsenal are also all but unknown, though it seems possible that at least a few were assembled from old Chuo Kogyō components as the military situation became increasingly desperate in 1944–45.

Combat experience showed the Type 96 to be acceptably reliable, largely because the troublesome lubricator was built into individual magazines and not the gun itself. Many were turned against their erstwhile owners by US and Australian captors, who generally appreciated the mild recoil of the 6.5×50mmSR cartridge – compared with .303in or .30-06 – which helped to promote accuracy even when burst-firing.

HOTCHKISS AND THE SPANISH CIVIL WAR

The Spanish Army, though armed largely with Mauser rifles, seems to have accepted the Hotchkiss machine gun prior to 1914 as the Ametrallador Mo. 1903 and thereafter made guns under licence. After World War I ended, and lessons were being analysed by almost all military authorities (even those, such as Spain, that had remained neutral), the M1922-type Hotchkiss was selected as a *Fusil Ametrallador* or 'machine rifle', which could be issued to infantry units to provide fire support.

A manufacturing licence was duly negotiated with Hotchkiss & Cie and work on what became known as the Fusil Ametrallador Mo. 1922/25 began in the state-owned Fábrica de Armas de Oviedo. Several thousand guns had been made by the early 1930s.

In July 1936, however, the Spanish Civil War began. Though often characterized simply as a Soviet-backed alliance of anarchists, communists and socialists fighting fascists backed by Germany and Italy, the war was much more complex – overshadowed by World War II, to some extent hiding bloody massacres and dreadful casualties (though the aerial bombing of Guernica on 26 April 1937 is usually over-dramatized).

Spain proved to be a testing ground for many new weapons. In addition to Spanish forces, the participation of the German Legion Condor and its Italian equivalent, the Corpo Truppe Volontarie, the International Brigades and a surprising number of foreign fighters took up arms on both sides. Weapons were in such short supply that many countries, despite signing a non-involvement agreement, took the chance to empty their storerooms. Almost anything that would shoot could be included. Consequently, though there were many modern rifles and machine guns among consignments sent to Spain from Germany, Italy, Poland and the Soviet Union, there were also old French Kropatschek and Italian Vetterli rifles chambered for large-calibre cartridges loaded with black powder. The deliveries included at least a few 7×57mm Hotchkiss machine guns, said to have been supplied from pro-Nationalist Mexico, and some 8×51mmR guns from France and Poland.

The Oviedo factory stood in Republican-held territory when war began, but the local military commander, Colonel Antonio Aranda, sided with the Nationalists. Consequently, the guns that languished in store were appropriated and much of the production machinery was evacuated to a new site in La Coruña. Here, in addition to Mauser rifles and Spanish-made and reconstructed French 7×57mm M1914 Hotchkisses, a simplified version of the Mo. 1922/25 was made as the Fusil Ametrallador Mo. 1925/38 or 'FA Oviedo-Coruña'. Work continued after the Civil War ended in April 1939 with a Nationalist victory, and was still under way when, in 1941, the first guns derived from the Czechoslovakian ZB vz. 30 were completed in the rebuilt Oviedo manufactory. The Mo. 25/38, a somewhat simplified version of the Mo. 25, fed from a top-mounted box magazine instead of the side-mounted strip and had a short wooden foregrip.

The quantities involved are disputed. The Republicans are said to have kept 628 M1914-type Hotchkiss machine guns and about 1,000 of the 1922-pattern machine rifles, while the Nationalists seized 1,400 and at least 2,500 respectively. In addition, a small number of original Hotchkiss Portatives, presumably chambering the French 8×51mmR cartridge, arrived in Spain aboard the steamship *Campeche* on 4 October 1936. These are said to have been consigned from the Soviet Union, where they had presumably been kept since arriving in Russia prior to the October Revolution (7–8 November 1917) as part of the French aid programme that included the archaic Kropatscheks (Heinz 2016: 14–15).

THE JAPANESE TYPE 99

Prompted by combat experience in Manchuria, with its extremes of temperature, and by the issue of the 7.7mm Type 99 infantry rifle, Japanese experiments soon began to develop a large-calibre version of the Type 96 light machine gun. Chambered for a new 7.7×58mm rimless cartridge, the Type 99 light machine gun (*Kuku shiki keikikanjū*) resembled the Type 96 Hotchkiss-derivative externally, though there were differences in the machining of the receiver and in the design of the barrel latch. Headspace was adjustable, overcoming a particular weakness of the

Hitachi Hieki-made Type 99 light machine gun no. 9151 dates from November 1943.
(Morphy Auctions, www.morphyauctions.com)

Markings on a Type 99. Unreadable though these may seem to those of us unfamiliar with a syllabic alphabet, interpretation is straightforward. The mark shown here contains the piled-cannonball mark of the supervising arsenal (Kokura) ahead of the trademark of Hitachi Heiki KK, the actual manufacturer, and then the designation ku ku shiki ('99 Type'). The next line contains the sequential serial number 452, prefixed by the encircled hiragana symbol *re*. The significance of these prefixes, usually associated only with Hitachi products, remains unknown. Arisaka rifles use hiragana prefixes taken from the poem *Iroha* to distinguish blocks of 100,000, but Japanese machine guns were not made in such huge quantities. Principal inspectors' marks seem more plausible, but this method does not seem to have been shared by other Japanese military firearms. The third line contains the date '17.2' prefixed by a character representing the nengō or imperial reign period – in this case 昭, shō counting from 1926 as '1'. (Only three reign periods are relevant to Japanese Hotchkiss and Hotchkiss-derivative machine guns: 明Meiji (1868–1912), 大 Taishō (1912–26) and 昭 Shōwa (1926–89).) The gun, therefore, was made in the second month of the 17th year of the Shōwa nengō: February 1943. A string of four small marks, applied by inspectors, is also present. (Morphy Auctions, www.morphyauctions.com)

earlier pattern, and the troublesome lubrication system was abandoned. The new magazine was not so sharply curved as its predecessor and a monopod could be fitted beneath the butt.

Made by Chuo Kogyō (inspected by Nagoya Army Arsenal), by Hitachi Heiki (inspected by Kokura Army Arsenal), by Kokura Army Arsenal itself and also by Hōten Zoheishō, the Type 99 was 1,181mm long, had a 545mm barrel, and weighed 10.4kg with its bipod. The detachable box magazine held 30 rounds, muzzle velocity was 715m/sec with Type 99 ball ammunition, and cyclic rate was 700–800rd/min. A 2.5× Type 96 telescope sight could be attached to the left side of the body and an armour-plate shield for use in defensive positions was available when needed.

Total production of the Type 99 has been estimated as 53,000 guns, though details are lacking. Hitachi guns 452 and 18094 were made in February 1942 and December 1944 respectively. Kokura guns 127 and 1338 date from July and October 1942 respectively. Nagoya production did not begin until 1943; gun no. 1903 was made in September 1943 and no. 9157 dates from May 1944. Hōten's output has been estimated at 13,000, but evidence is lacking.

The manufacturing quality of Type 99 guns declined appreciably as the war ran its course. Chroming of the bore and gas-regulator chamber was abandoned early in 1944 to save valuable raw material, the monopod was omitted on many guns completed after October 1944, and the rear sight was simplified in November that year. The optical-sight base was eliminated in May 1945, serial numbers disappeared from the lesser parts, and German MG 15 magazines, displaced from air service, were adapted to fit. Analysis of individual late-war Type 99 guns often reveals shortcuts taken unofficially at a time when aircraft guns were accorded priority.

The Type 99 provided the basis for the Type 1 Model 1, a paratrooper's gun developed by Nagoya Army Arsenal. Made only in small numbers, the Type 1 had a characteristic hollow steel pistol-grip that could be folded forward to protect the trigger and trigger guard. The magazine and the barrel could be readily separated from the receiver, to fit into a paratrooper's leg pouch, but these sub-assemblies were still heavy and cumbersome.

HOTCHKISS: THE FINAL ACTS

The M1914 Hotchkiss was still the preferred French front-line machine gun in 1940, serving with the Vichy and Free French forces, the German-raised paramilitary *Milice*, and throughout the French colonial empire.

Most surviving M1914 guns had received the 2,400m-range dial-elevated sights calibrated for the 8mm Balle 1932 N and were also sometimes accompanied by the M1928 anti-aircraft adaptor, which fitted onto the tripod to elevate the gun to shoulder height.

Some British-made light Hotchkisses, which had been redesignated Guns, Machine, Hotchkiss, No. 2 Mk I* in 1926, were still being carried aboard armoured vehicles when World War II began. Others, remaining in store, were refurbished for emergency service once Bren Guns were lost in large numbers at Dunkirk in May–June 1940. It is also possible that survivors of trials undertaken in 1935–36 with 35 13.2×99mm Model 30 Hotchkiss machine guns, were also pressed into British service. As supplies of new Bren Guns increased, however, some of the No. 2 Hotchkisses were passed to the imperial navies and the mercantile marine. Others served with British, Australian and Indian units involved in North Africa and the Mediterranean theatre, usually for local anti-aircraft defence, while Hotchkisses captured by German forces in Crete, for example, served the Wehrmacht under their *Fremdengeräte* designation Maschinengewehr 136(e) alongside essentially similar '136(g)' machine guns taken from the Greeks.

However, most of the old British guns saw out their active lives with the Home Guard. At least 10,993 of them had been refurbished, principally by Westley Richards & Co. Ltd in Birmingham (4,170), the Royal Small Arms Factory in Enfield (4,000), BSA Guns Ltd in Birmingham (1,580), and J. Boss & Co. in London (1,093) (Skennerton 1988: 60). The Hotchkiss No. 2 was declared obsolete in June 1946.

The Japanese continued to use their Hotchkiss machine guns and adaptations thereof until the Pacific War ended on 15 August 1945, but many Chinese Hotchkisses – ex-French M1914 and Chinese copies – survived not only World War II but also the conflicts in Korea (1950–53), French Indo-China (1945–54) and Vietnam (1955–75). At least one Turkish M26 Hotchkiss mitrailleuse légère came back from Afghanistan in 2016, so who knows what could still be out there!

Men of 1st Battalion, The King's Own Scottish Borderers, clearing the streets of Caen on 10 July 1944, make good use of a captured M1914 Hotchkiss. (Author's Collection)

IMPACT
A ground-breaking weapon

ASSESSING THE HOTCHKISS

The Maxim and Hotchkiss guns performed reliably during the Russo-Japanese War, and were retained after that conflict had ended. The Russian Madsens were overlooked, but that this was due largely to internal politics – the Imperial Russian Army authorities blamed much of their defeat on the cowardice of the cavalry – went unseen.

Consequently, many Western observers assumed that only heavyweight machine guns had a future. Maxims could fire tens of thousands of rounds in trials with very few jams, though, in the early days, problems arose from changes in ambient conditions; from the variable pressures generated by differing batches of ammunition, often arising from lack of experience in the manufacture of smokeless propellant; from inappropriate materials used in the construction of key components; from poor manufacture; and, sometimes most importantly of all, from almost no training in the use of the guns themselves.

The Hotchkiss was much simpler than the Maxim, but trials showed that it was incapable of sustaining fire for lengthy periods. Even though the barrel was fitted with radiator fins, it soon heated up to a point where (if rapid fire was perpetuated) bullets no longer engraved in the rifling and, ultimately, the bore was ruined. Most water-cooled guns could fire tens of thousands of rounds without stopping. Regulations suggested that the barrel of the M1914 Hotchkiss was to be swabbed with cold water after 1,000 rounds, and allowed to cool for approximately four minutes before fire should resume. The barrel was surprisingly easy to exchange, however, thus allowing fire to be maintained as long as spare barrels were to hand.

According to an article published in the *Journal of the Royal United Services Institute* in February 1910, the Japanese Hotchkisses that had played a vital role in the Battle of Hei-Kou-Tai (25–29 January 1905), jammed on average once in 300 shots and extractors broke on the scale of one per gun. One Hotchkiss had been put out of action when a cartridge exploded as it was being pushed from the feed strip into the chamber – whether a 'cook-off' or a defective primer was unreported – but together these guns contributed greatly to a Japanese victory that had cost the Russians 7,000 men.

A US machine-gun crew and observers training with an M1909 Benét-Mercié Machine Rifle prior to 1917. (US Army)

The participants in the Russo-Japanese War, which lasted for 19 months, faced a wide range of combat conditions. Temperatures plummeted as winter approached, and the performance of machine guns often deteriorated. Water in the jackets of the Russian Maxims froze, and both sides noted that as temperatures rose rapidly during continuous firing, jams became increasingly frequent and parts broke more often. Yet machine-gunners became increasingly adept at clearing jams as they gained combat experience, and there can be no doubt that the value of the machine gun only slightly diminished as the result of an occasional jam.

The Japanese soon realized that covering their guns with blankets minimized the effect of sub-zero temperatures, and, interestingly in view of the problems experienced by US Army machine-gunners during the Battle of Columbus, had even provided small battery-powered electric lights, equipped with effective shrouds, so that their gunners could load Hotchkiss feed strips more efficiently during night-fighting.

Another problem arose from ammunition supply, asking questions that quartermasters and supply columns often failed to answer. The Russian armies at the Battle of Mukden mustered 340,000 men, and there were about 270,000 Japanese who alone fired more than twenty million rifle and machine-gun rounds. The effective employment of machine guns is reflected in the horrendous casualties: 8,705 dead and 51,438 wounded Russians, and 15,892 dead and 59,612 wounded Japanese. Many Japanese commentators noted that the 'rat-a-tat' of the Hotchkisses was a great morale-booster, often encouraging men to surge forward in circumstances which were far from ideal. An officer of the 34th East Siberian Rifle Regiment took the same view, but from the Russian perspective:

In modern battles the harsh, broken rattle of the machine gun is heard uninterruptedly for whole hours, producing a disheartening and irritating effect on the men. In addition to the losses suffered by a detachment coming under fire of machine guns the enormous losses incurred in a brief period of time cause great depression. The greatest effect is produced, both morally and physically. It is not surprising, therefore, that the machine guns were christened by the men 'the devil's spout' … (Soloviev 1906: 33)

Yet many European armies, despite the positive assessments of their own military observers, often chose simply to ignore the lessons the large-scale use of machine guns had shown so clearly. There were several reasons for this lax and ultimate costly attitude.

Though machine guns had been brought to a satisfactory state of perfection, they were costly and complicated. They performed well enough in the hands of trained men – though rarely as well as they did in trials – but had proved vulnerable to counter-attack. Some observers, noticing how Russian Maxims in particular had been lost to the Japanese, worried that machine guns could be turned against their erstwhile owners – creating an obsession not so much with battlefield mobility but more on the ease with which machine guns could be withdrawn from the battlefield if threatened.

The result was initially to confine machine guns to fortresses, strongpoints and similar defensive positions, or to classify them as light artillery. Cumbersome wheeled carriages were retained, often requiring the services of two or more horses to move them, and vulnerability to counter-fire increased by raising the bore-line far too far from ground level.

The French had made mistakes of this type with the de Reffye Mitrailleuse during the Franco-Prussian War, even though these primitive machine guns had shown their value when used as close-range infantry support weapons, but many people were still following the same tack as late as 1914.

Some armies had formed special infantry-support units, usually in the form of independent machine-gun companies, but their effectiveness was restricted by the small numbers of guns that were purchased. World War I was soon to show that maximum 'fire density', which meant quantity (if necessary at the expense of quality), was the most important criterion; and fire density was to show just how vulnerable men were to machine-gun fire if their commanders had no appreciation of the weapons' capabilities.

War in the Pacific (opposite)

Advancing into the field of fire of machine guns manned by Japanese defenders, US Marines emerge from an LCP or 'Higgins Boat' to assault a Japanese-held island in the Kwajalein archipelago in 1944. The defenders have a 7.7mm Type 92 machine gun and a 7.7mm Type 99 light machine gun. They wear khaki uniforms, their helmets with net covers and foliage tucked into the mesh as camouflage. The gunner has a belt, haversack and water bottle; the loader, waiting to feed the new cartridge strip he has just taken from the box between the two men, carries the original two-buckle ammunition bag over his shoulder. The light-machine-gunner has a back pack, with entrenching tool and rice cooker. The Type 92 was once protected by a roofed-over bunker, but a shell from a US Navy warship – visible on the horizon – taking part in fire-support bombardment has reduced it to a fire-blackened log wall. An ejected case is still in mid-air to the right of the gun, and the surround is littered not only with the cardboard cartons in which the strips were supplied but also by expended strips which have fallen out of the right side of the breech.

The opening stages of World War I seemed to confirm that the water-cooled recoil-operated Maxim and its near-relation, the Vickers Gun, maintained fire better than the air-cooled gas-operated Hotchkiss. This fact was rarely disputed, though the air-cooled guns, which were lighter and in no way tied to ready supplies of coolant to replace that which the heat generated on firing had evaporated, could be manoeuvred more easily. This sometimes suited them to tactics in which fluidity was important.

If the M1914-type Hotchkiss was an evolutionary dead end as far as sustained-fire machine guns were concerned, the Portative, despite several inherent shortcomings, pointed the way to the future. Though the recoil-operated M1917 Browning was a runaway success, post-war experimentation tended to focus on gas operation. Recoil-operated guns were usually more complicated, and, therefore, more difficult and more expensive to manufacture. Gas-operated guns were in essence very simple, with few and relatively lightweight moving parts; their weaknesses tended to centre on the ease with which they could handle variations in pressure generated in individual cartridges and the accumulation of propellant fouling.

The shape of the cartridge case, tapered or parallel-sided, also tended to have more influence on the performance of gas-operated weapons than in recoil-operated types. The almost parallel-sided Japanese 6.5×50mmSR Arisaka cartridge performed very differently to the 8×51mmR cartridge for which the Portative had been developed. The body of the French cartridge tapered considerably, and the rim was exceptionally sturdy. When the gun fired and extraction began, the French spent cartridge case started back out of the chamber more easily than its Japanese equivalent in which residual pressure tended to push the cartridge case momentarily outward against the chamber wall. Consequently, the Japanese extractor sometimes jumped out or even tore through the groove in the case head whereas the French equivalent, with a robust projecting rim to grasp, rarely suffered the same problems.

The Portative was not the first successful light machine gun, as the Danish Madsen had been introduced as early as 1902 in what was substantially its perfected form. Madsens had sold in quantity prior to 1914, and were used by the Allied and Central Powers forces alike during World War I. The Germans acquired batches of Madsens to equip *Gebirgsjäger* (mountain troops), the Russian requirement for Madsens was sufficient to encourage Dansk Rekylriffel Syndikat to begin constructing a machine-gun factory in the Russian town of Kovrov, and even the British placed a small order. The Madsen had a box magazine on top of the receiver, feeding downwards under a combination of spring and gravity, and had proved capable of handling a variety of rimmed and rimless cartridges. Its biggest problem was the lack of the primary extraction inherent in most rotating-bolt designs, though the mechanism – a clever adaptation of the pivoting Peabody breech block – was both sturdy and reliable.

Anglo-American sources often favour not Madsen but Colonel Isaac N. Lewis, a US Army artilleryman who had had a long and acrimonious battle with the US Army Chief of Ordnance, Brigadier General William Crozier. Consequently, progress was delayed until a manufacturing licence was concluded in Belgium, and success was ensured only when the British Army ordered huge quantities of Lewis Guns from BSA Guns Ltd in 1915.

The German MG 08/15, a cumbersome water-cooled 'portable' variant of the standard Maxim MG 08, was nothing other than an expedient. The problem was tactical: until the value of *Sturmtruppen* could be verified, efforts were almost always centred on the desire to sustain fire instead of focusing on portability.

This Austro-Hungarian machine-gunner sits behind a captured M1914 Hotchkiss on the standard tripod mount; note the ammunition belt, which is actually made of articulated three-round striplets, and the winding handle extended on the right side of the ammunition box. (Author's Collection)

The Hotchkiss Portative was undoubtedly the first gas-operated machine gun to succeed in an airborne role, at least prior to the arrival of stripped-down Lewis Guns. Most of its rivals, such as the Maxim and the Vickers guns, were recoil operated. Only the M1895 Colt–Browning could be cited as a rival, though its gas-propelled pivoting lever, the so-called 'Potato Digger', was ill-suited to use in the air unless a cumbersome protective cage or guard was fitted beneath the barrel. The Colt was developed by Marlin Firearms Company into an effectual aeroplane and armoured-vehicle weapon, but only after extensive changes made to the gas system had eliminated the swinging lever.

NEW DEVELOPMENTS

After World War I had ended, though millions of weapons were destroyed, most armies took the opportunity to analyse the lessons of war. One obvious advance had been the introduction of light automatic weapons which, though incapable of replacing heavy water-cooled sustained-fire machine guns, had the great advantage of portability. The French tactic of 'assault-at-the-walk', widely disparaged in the trench-locked opening phases of World War I, was seen to have advantages once the Lewis Gun and the Browning Automatic Rifle became available.

The performance of the French machine guns during World War I left much to be desired, particularly when compared with the German Maxim and the British Vickers guns. Though economics dictated that the M1907/16 T Saint-Étienne should be retained, the M1914 Hotchkiss had proved to be the more reliable machine gun and so was standardized instead. Work also began to develop a new light machine gun.

Among the first of the new gas-operated guns developed to challenge the recoil-operated Madsen and the gas-operated Hotchkiss Portative and Lewis Gun was the Berthier, tested by the US Army in 1917 and subsequently exploited by Vickers but with pre-1914 origins; cam fingers on the piston-rod extension acted on lugs on the breech block to raise the tail of the block into the roof of the receiver. A specially hardened steel bar in the receiver, running transversely, minimized wear. The

Two views of a Brazilian 7×57mm Type II Hotchkiss Fusil Mitrailleur, with bipod. (Morphy Auctions, www.morphyauctions.com)

Czechoslovakian Praga and its derivatives (ultimately including the Bren Gun) were similar, but simpler and sturdier.

Trials were undertaken in France during the early 1920s with the Madsen, the light M1922 Hotchkiss, the Lewis Gun, the Browning Automatic Rifle and the Berthier. These were joined by the MAS 22 (a copy of the BAR) and then in April 1923 by the MAC 23. The MAC 23 soon outperformed the BAR and the MAS 22, and the Berthier, even though it had been designed by a Frenchman, was rejected by a government reluctant to pay royalties to Vickers-Armstrong.

Hotchkiss had developed an improved version of the Portative, perhaps aimed at the North American market, but the advent of war in 1914 inhibited progress. The patent (INPI 1913) includes illustrations of a bipod combined with a scabbard for the standard US M1905 sword-bayonet and the bayonet attached beneath the barrel – presaging the Japanese Type 96.

The M1922-type Hotchkiss was a modernized M1914 action, retaining the familiar flap lock but with pinned links instead of a cam-type lifter. Patents were granted in 1923–25 to protect the firing mechanism, the feed being either a belt of jointed metallic striplets or a detachable box magazine on top of the receiver: the former was classed as a *mitrailleuse légère* ('ML': light machine gun), while the latter was a *fusil mitrailleur* ('FM': machine rifle). Barrels could be plain or finned.

The similar but unattributable Type III no. 16239, fitted with a socket for a pintle mount. (Photo courtesy of Cowan's Auctions Inc., Cincinnati, OH)

The guns had distinctive pistol-grip butts, bipods with noticeable 'rocker' feet, and flash-hiders that were cut obliquely to serve as compensators. The cocking slide lay on the right side of the breech. Advertised as the 'M1924', 'M1926 or 'M1934', they were exported to Greece and Romania (6.5×53mm); Brazil, Chile, the Dominican Republic and Spain (7×57mm); Lebanon (7.5×58mm); Turkey (7.65×53mm); and Greece and China (7.9×57mm). Some of the Greek Hotchkisses will be

The Hotchkiss M1922 aircraft gun, from a manufacturer's handbook. (Author's Collection)

found with Italian Breda-type bipods and an optical-sight bracket on the left side of the body, though the sights are generally missing.

A few 0.303in M1922 adaptations were tested unsuccessfully in Britain against the Vickers-Berthier, and about a thousand 7.92×57mm guns went to Czechoslovakia for field trials against a similar quantity of Darne machine guns and the Praga I-23, prototype of the ZB vz. 26. Derivatives included single- and double-barrel aircraft guns, which usually fed from an improved cartridge belt patented in 1925.

The French purchased a few Mitrailleuses Légère Hotchkiss Modèle 1934 light machine guns, chambered for the 7.5×54mm rimless cartridge, for service in Indo-China. The M1934 was 1,216mm long, had a 600mm barrel, and weighed 9.58kg with the lightweight bipod attached. The action fed a 30-round metal strip from the right, muzzle velocity was 860m/sec with the 7.5mm Balle 1933 C, and cyclic rate was approximately 450rd/min.

But the French authorities, just as they had done in 1905 and again in 1907, had already rejected the Hotchkiss in favour of a new gas-operated gun developed by government technicians. Adopted as the Fusil Mitrailleur Modèle 1924, and introduced to service in 1926, it had a distinctive double trigger and a short wooden fore-end. The action was an adaptation of the Hotchkiss system, relying on a swinging Colt–Browning link to tip the tail of the bolt up against a recess in the receiver. A top-mounted box magazine was fitted, the only quirky feature being the double-trigger firing system. The front trigger, which embodied a disconnector, gave single shots; the rear trigger, which held the sear and sear buffer down, allowed automatic fire.

No sooner had production of the M1924 begun, however, than problems with the new 7.5mm rimless cartridge arose. The dimensions of the case were so similar to those of the German 7.9mm cartridge that accidents were feared, and so a decision was taken to shorten the 7.5mm version by 4mm. The result was the 7.5mm Balle 1929 C, and the M1924 guns that had been chambered for the original-length cartridge were returned to the ordnance factories for revision.

Many M1924/29 guns were rifled and sighted for M1933-type heavy-bullet ammunition, which performed far better at long-range than the standard type; these guns have barrels marked with a large 'D'. The M1924 remained in production until the end of World War II and lasted in front-line service into the 1960s.

A typical 7.5mm Fusil Mitrailleur M1924/29. Chambered for the 7.5×54mm rimless cartridge, the M1924 was 1,070mm long, had a 500mm barrel, and weighed 8.93kg without the detachable 25-round box magazine or bipod. Cyclic rate was approximately 450rd/min. The M1924 and its M1924/29 successor proved to be sturdy and efficient, and was an immense improvement on the Chauchat if not quite such an advance on the Portative. However, it was unnecessarily complicated with 164 components in the gun alone, including an assortment of pins and no fewer than 21 springs; by contrast the Soviet DP had only 53 components. The M1924 was easily field-stripped, but the barrel-change system was poor. Not only did the barrel screw into the receiver, but it carried the bipod and lacked a handle. In addition, the gas-cylinder tube had to be detached before the barrel unit could be removed. (Rock Island Auctions, www.rockislandauction.com)

Hotchkiss & Cie also introduced a heavy machine gun, inspired by the Balloon Gun of World War I. The M1930 was an enlarged and modernized form of the M1914, intended for a variety of uses. Locked by the well-tried flap system and chambering a 13.2×99mm cartridge giving a muzzle velocity of approximately 800m/sec with standard ball rounds, the standard tripod-mounted infantry-pattern M1930 had a single spade grip and a barrel that was finned in its entirety, though there were three different diameters. The gun was approximately 1,660mm long, had a 1,000mm barrel, and weighed 37.5–39.7kg depending on pattern. Feed was from either a 15-round metal strip or a 25-round detachable box magazine. Cyclic rate was approximately 450–550rd/min.

Mounts for the M1930 ranged from sturdy tripods to wheeled and shielded mounts for heavy support and anti-tank roles; pillar mounts were offered for anti-aircraft defence aboard small warships. Some M1930 guns were also used in the air. The French forces used 13.2mm Hotchkisses in fortresses, on armoured vehicles, and aboard warships. Others were sold to Chile, China, Mexico and Romania. They were also made under licence by Breda in Italy (as the 'Mo. 1931').

After two years of negotiations, begun in April 1927, the Kokura Army Arsenal (trading as the 'Kokura Industrial Company') secured licensing rights to the M1930-type Hotchkiss machine gun, which was introduced into Japanese service in 1933–34 as the Type 93 heavy machine gun. This chambered a 13.2×99mm cartridge and fed from a top-mounted spring-loaded box magazine holding 30 rounds. Originally intended as an anti-aircraft weapon, issued with special sights and mount, the Type 93 was also used for infantry support and anti-tank warfare.

Unfortunately the 13.2×99mm cartridge had the same case-length as the 0.5in Browning cartridge (12.7×99mm), and so, owing to the potentially fatal chance of accidentally using the wrong ammunition, Hotchkiss decided to shorten the case to 96mm. The M1935 was a minor variant of the original gun, chambering the reduced-length cartridge. The change in length was small enough to obviate the necessity for wholesale changes in the gun, but World War II intervened before anything could be done to encourage sales internationally.

CONCLUSION

The Hotchkiss was the first machine gun to prove the value of gas operation, being both simpler and less costly than recoil-operated designs. It promoted mobility – if observers and military authorities could be convinced – and to some extent freed the machine gun from largely static roles. This was not obvious until World War I proved conclusively that the horseman was no match for the machine gun, and also that automatic weapons could be adapted to a wide variety of uses, but the advent of lightweight weapons such as the Hotchkiss Portative and the Lewis Gun presaged the demise of recoil operation. Browning, Maxim and Vickers guns soldiered on until the end of World War II, but virtually all the designs developed since 1945 have been gas operated. Light machine guns and automatic rifles have relied on gas operation since the 1920s.

Of course, the Hotchkiss was not without its problems. When it first appeared, smokeless propellant was comparatively new and the manufacturing processes were still being refined. Pressures generated in cartridges as they were fired sometimes were insufficient to complete the reloading cycle; at other times, occasional pressure-spikes opened the breech too quickly, and could cause damage to components.

Hotchkiss guns worked well with rimmed taper-body cartridges such as the French 8×51mmR or British .303in, which extracted well, but rimless ammunition sometimes presented problems if the cartridges were parallel-sided. These cases tended to stick to the chamber walls for a micro-second longer than tapering designs, which usually began to move back almost as soon as the breech opened. The earliest Hotchkiss guns were offered in light and heavy patterns, the former with a plain cylindrical barrel and the latter with four large-diameter fins intended to present a greater surface area to the atmosphere and thus enhance cooling.

The Hotchkiss had always been and would remain air-cooled in pursuit of simplicity, but even though its barrels became heavier and more

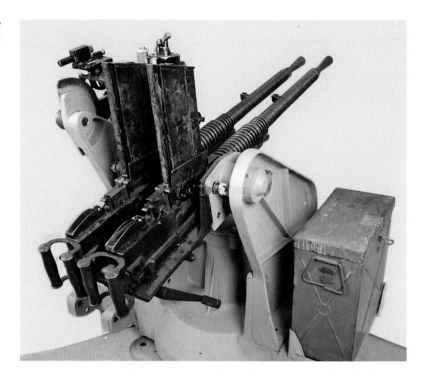

fins were added to increase the surface area presented to cooling air (or even a bucket of water), air-cooled guns could never sustain fire as effectively as their water-cooled rivals in land service. In the air, things were very different, but in that operational environment the Hotchkiss was handicapped by its quirky strip feed.

In combat, however, overheating does not seem to have been as much of a handicap as the Hotchkiss machine gun's detractors claimed. It seems that experienced machine-gun crews could fire the M1914 for lengthy periods if the rate of fire was restricted to no more than four strips per minute (120 rounds). The barrel soon glowed red, but a point of equilibrium had been reached: the barrel dissipated heat at much the same rate as it was being generated, and so did not sustain serious damage.

The Hotchkiss was also one of the first machine guns to sell internationally on a large scale: only the Maxim sold more widely, and in greater numbers, prior to 1914. The performance of the Hotchkiss in World War I was creditable, but it was overshadowed by the Brownings by the time the Armistice was signed. Though developed into an effectual light machine gun in the early 1920s, the Hotchkiss was overlooked once again by the parsimonious French government and denied the support from which some if its rivals benefited. Though substantial quantities of rifle- and large-calibre machine guns were sold in the 1920s and 1930s, the arms-making division of Hotchkiss & Cie was nationalized by the left-wing government of the Front Populaire in 1936. Together with worldwide economic depression and the waning influence of France, both politically and industrially, this would have brought work to an end had not World War II done so first.

BIBLIOGRAPHY

Allgemeine Kriegsdepartement (1916). *Beute-Maschinengewehr*. Berlin: Reichsdruckerei.

Buigné, Jean-Jacques, and Jarlier, Pierre (2001). *Le "Qui est qui" de l'arme en France de 1350 à 1970*. La Tour du Pin: Editions du Portail.

Ewen, Murray (2017). 'The Gallipoli Maxims Revisited', in *The Gallipolian* No. 145.

Federov, Vladimir (1939). *Evolyutsiya strelkovogo oruzhiya, Chast II, Razvitie avtomaticheskogo oruzhiya*. Moscow: Voenizdat.

Hamilton, Lt-Gen. Sir Ian, KCB (1907). *A Staff Officer's Scrap-Book During the Russo–Japanese War*. Volume 2. New York, NY: Longmans, Green & Co.

Hamilton, John (2015). *Gallipoli Sniper. The Remarkable Life of Billy Sing*. Barnsley: Frontline.

Heer, H. & Naumann, K. (2000). *War of Extermination: The German Military in World War II, 1941–44*. New York, NY: Berghahn Books.

Heinz, Leonard R. (2016). 'Small Arms of the Spanish Civil War', accessed from <https://www.forgottenweapons.com>.

Hughes, James B., Jr. (1968). *Mexican Military Arms. The Cartridge Period 1866–1967*. Houston, TX: Deep River Armory.

Huon, Jean (1979). *Un siècle d'Armement Mondial, armes à feu d'infanterie de petit calibre*, tome 3 [Finland-Italy]. Paris: Crepin-Leblond.

INPI (1907). French Patent 375307, *Dispositif permettant le démontage et le remontage rapides du canon sur la boîte de culasse des mitrailleuses Hotchkiss et autres*, accepted 10 May 1907.

INPI (1908). French Patent 391600, *Dispositif de pointage en hauteur, pour mitrailleuses*, sought 24 June 1908.

INPI (1912). French Patent 438711, *Machine pour garnir de cartouches les bandes de chargement de mitrailleuses automatiques*, sought 6 January 1912.

INPI (1913). French Patent 4617882, *Perfectionnements apportés aux fusils automatiques*, accepted 6 November 1913.

INPI (1915). French Patent 502393, *Bande de chargement articulée pour mitrailleuses Hotchkiss*, sought 15 July 1915.

Král von Dobrá Voda, Adalbert Ritter (1904). *Der Adel von Böhmen, Mähren und Schlesien*. Prague: publisher unacknowledged.

Liddell Hart, B.H. (1959). *The Tanks. The History of the Royal Tank Regiment and its predecessors …* New York, NY: Prager.

Longstaff, Major F.V. & Atteridge, A. Hilliard (1917). *The Book of the Machine Gun*. London: Hugh Rees.

Martin, Colonel en Retraite Jean (1974). *Arms à feu de l'Armée Française 1860 à 1940*. Paris: Crepin-Leblond.

Scarlata, Paul (2019). 'Model 1909 Benet-Mercie Machine Rifle: U.S. Army's First Light Machine Gun', www.firearmsnews.com, 7 January 2019.

Skennerton, Ian D. (1988). *British Small Arms of World War 2: the complete reference guide to weapons, makers' codes & 1936-1946 contracts*. Margate, Queensland: self-published.

Soloviev, L.Z. (1906). *Actual Experiences in War: Battle Action of the Infantry; Impressions of a Company Commander*. Washington, DC: US Government Printing Office.

War Office, Britain (1908). *The Russo-Japanese War: Reports from British Officers Attached to the Japanese and Russian Forces in the Field*. London: His Majesty's Stationery Office.

Woodman, Harry (1989). *Early Aircraft Armament. The Aeroplane and the Gun up to 1918*. London: Arms & Armour Press.

INDEX